Broadband Networks

A Manager's Guide

Robert P. Davidson

WILEY COMPUTER PUBLISHING

John Wiley & Sons, Inc.
New   York • Chichester • Brisbane • Toronto • Singapore

Publisher: Katherine Schowalter
Editor: Robert Elliott
Managing Editor: Robert S. Aronds
Text Design & Composition: Publishers' Design and Production Services, Inc.

Designations used by companies to distinguish their products are often claimed as trademarks. In all instances where John Wiley & Sons, Inc. is aware of a claim, the product names appear in initial capital or all capital letters. Readers, however, should contact the appropriate companies for more complete information regarding trademarks and registration.

This text is printed on acid-free paper.

This publication is designed to provide accurate and authoritative information in regard to the subject matter covered. It is sold with the understanding that the publisher is not engaged in rendering legal, accounting, or other professional service. If legal advice or other expert assistance is required, the services of a competent professional person should be sought.

Library of Congress Cataloging-in-Publication Data:

Davidson, Robert P.
    Broadband networks: a manager's guide / Robert Davidson.
        p.   cm.
    Includes index.
    ISBN 0-471-13885-1 (pbk.: alk. paper)
    1. Broadband communication systems.   I. Title.
    TK5103.4.038   1996
    004.6—dc20   95-49740
                                                                            CIP

Printed in the United States of America

10 9 8 7 6 5 4 3 2 1

To my wife Myrna,
and my sons Alan, Brian, and Kevin

Special thanks to Barbara Sussman Goldberg for her assistance

Contents

SECTION 2   LANs TO WANs

5   LANs   73

6   LAN Internetworking   99

SECTION 4: STRATEGIC CONSIDERATIONS

11 Planning Considerations   217

Introduction

Rapid advancements in communications technology are changing society in ways that were only dreamed of a scant two decades ago. The globalization of economies, the tying together of personal computers into powerful electronic facilities, and the sending of mail in the form of facsimile are very recent achievements, yet they are being accomplished via the public telephone network. To meet these needs, the network is continuously adapting to the ongoing conversion of information to digital electronic signals that are capable of being transported over telephone lines. To put this in perspective, however, it is estimated that 97 percent of our information is still on paper or microfilm, implying that the transference to digital data is only just beginning.

The onslaught of multimedia information—voice, video, and image—has already transformed the bulk of public network traffic from voice to data. The scaling down of information systems has created an equally rapid and often turbulent growth of local area networks (LANs). Originally, LANs were put together—often with no central planning—for the sole purpose of addressing very local departmental issues. Today, these once separate islands of computers are so extensive and internetworked to such a degree that it is difficult to tell where the LAN ends and the wide-area network (WAN) begins. Compounding the current situation is a host of communication technologies such as voice processing, cell and frame relay, asynchronous transfer mode (ATM), and synchronous optical network (SONET), all of which are encouraging the widespread use of the so-called "killer applications." This new array of software supports mobile computing, image processing, video conferencing, executive information systems, groupware, client/server databases, and voice mail. The broadband networks and services that support these applications—switched multimegabit data service (SMDS), broadband integrated services data network (BISDN), ATM, fiber distributed data interface (FDDI), fast Ethernet, and so forth—are, in turn, all based upon the ubiquitous fiber cable. As a result of this proliferation of technologies and services, knowing how to use them as well as how to expand the bandwidth of both private and public networks has become crucial to information management and, consequently, to competitive enterprise.

In the global business arena as elsewhere, the fittest, that is to say, the most adaptable and appropriate, will survive. The mainframe has fallen prey to the personal computer, the proprietary computer network has yielded to the LAN, and now private corporate networks are giving way to integrated public network services. The need is for nothing less than desktop-to-desktop conductivity. Today's commerce demands the real-time transmittal of information to individuals as well as to entire corporations—when, where, and in whatever form it is needed. Information once available only to the power broker has become democratized. The necessity to collect, manage, and use globally available information has never been greater. In today's information-intensive industries, it is the ability to grasp knowledge quickly that provides the competitive edge; it is a given that information has become as important to a business's success as its capital assets. Few can afford to ignore this trend because the need for industries to be innovative continues to rise seemingly without end. Companies that adapt to the new information age will prosper because their ability to collect and manage information in real-time will spell the difference between success and failure.

Educating individuals to manage both equipment and people through this revolution in digital communications is also becoming critical to a company's success, since entire new generations of network technologies are being combined in various ways to replace traditional public and private networks, computers, telephones, and televisions. Many managers lack the adequate technical background to deal with the advanced networks of computers that have become vital to daily business. Often they must plunge into jobs without sufficient time for training. Worse, the underlying technological premise is changing so rapidly that even specialists have trouble keeping pace. In a few short years, decades of voice dominance in the public telephone network have given way to data; copper wire has ceded to fiber; and the separation between computers and their networks has become blurred. All of this indicates a pointed need for the proper person(s) to take charge of ongoing information technology endeavors.

This book is an attempt to provide the necessary foundation for the understanding of broadband networks. It details the latest fundamental changes taking place in the field with regard to technology, business orientation, and the marketplace and explores the numerous factors driving the development and deployment of broadband networks (Figure I1). At the same time that these networks are being deployed, however, the jury remains undecided as to which broadband network standards, equipment, and technologies will be accepted. It takes knowledgeable managers working closely with system engineers to plan

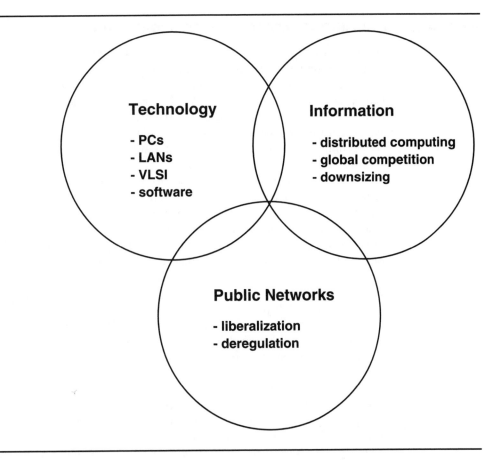

FIGURE I.1   Broadband network drivers.

and execute a migration. Keeping in mind that the business applications them-selves will ultimately determine the direction in which the network goes, this book provides a step-by-step road map for delivering a network that meets your company's needs.

Section 1 details the forces that are driving businesses toward broadband applications, providing valuable insight into how they could be used to make the business more competitive. Section 2 describes LANs and WANs, focusing on their design, implementation, and effective use. Section 3 tells all about broadband network technologies, particularly with regard to when and how they should be applied. Finally, this treatment would not be complete without a way to justify the investment. Section 4 provides the strategic considerations for the development and improvement of a corporate network.

SECTION 1: WHY BUSINESSES ARE MOVING TO BROADBAND

Chapter 1, *The Business Case for Broadband Technology*, focuses on how networks are being reengineered to encourage people to move from compartmentalized to cross-functional lines. To get there, not only must customers understand their options for LANs as well as for WANs, but also providers must understand the market dynamics. The companies—the broadband data network consumers—that prosper will be the ones which solicit the combinations of features and services uniquely appropriate to their businesses. The vendors that prosper will be those who offer these combinations.

Chapter 2, *New Applications*, describes the business operations that are changing the workplace. Databases, word processing, and spreadsheets that were once reserved for individuals are now within the purview of the group. They are greatly increasing the productivity of new network applications such as facsimile, E-mail, imaging, groupware, and multimedia. All of this is presented within the context of the societal and economic pressures that are driving the computer and telecommunications industries in converging directions. At the same time, trends to expand the home work force (telecommuting) and legislative initiatives that promote the use of the telecommunications network are creating unrivaled opportunities for broadband technologies and services. World-spanning communications networks have removed distance as a criterion for effective contact between people, businesses, and nations. With the Internet and other such connections, people can communicate with each other as well as with computers anywhere and anytime over the public telephone network, creating a de facto global collaborative workplace. Eventually, all businesses will be connected electronically, sharing a wide variety of information and commercial applications and consequently redefining modern commerce.

Chapter 3, *New Computing Networks*, details the computer trends that are accelerating the introduction of broadband networks. In the past decade, data has supplanted voice as the predominant source of public network traffic. The increased numbers of personal computers and sophisticated software applications that make businesses more efficient and effective have contributed to this situation as organizations have come to realize that pockets of information must be interconnected. Workgroups can now share knowledge in the forms of images and video as well as text. Entire companies automatically create network traffic as they track customers, create conferences for collaborative problem–solving, and access libraries of policies, documentation, and news. Collaborative computing provides real-time linkages into a myriad of potential network resources, enabling users to create, secure, update, and maintain the information residing in intelligent network elements. This torrent of demands is deluging the public and private networks, and creating pressures for intelligent and concurrently managed information superhighways.

Chapter 4, *New Communications Media*, describes the fundamental changes in semiconductors, fiber optics, computer software, and wireless technology that are making broadband networks possible. Information is already being rapidly delivered by networks with more connections between them and with more services. The creation of ever faster networks is being encouraged by advancements in silicon integrated circuits, fiber optics, and computer software technology. The suppliers of the services are both traditional and new. While wireless communications providers continue to pose a threat to telephone company local loop service revenues, the Internet competes with the long-distance carrier business.

SECTION 2: LANs TO WANs

Chapter 5, *LANs*, describes how an early form of broadband network, the LAN, has become the dominant computing environment, replacing earlier mainframe and minicomputer networks. The efficient and economical sharing of information and equipment has impelled the emergence of LANs from the backwaters of computing to the forefront of networking. This chapter also reports on why it is becoming less likely for users to be concerned only with their immediate environment. Today, people want to communicate and share resources and information, independent of geographic constraints, thus encouraging the development of better interfaces between LANs and WANs, as well as higher data rates. In addition, multimedia applications are pushing Ethernet LANs to their limits. As LAN traffic continues to increase, network operators must look to faster technologies such as ATM, FDDI, 100VG-AnyLAN, and fast Ethernet rather than rely on workarounds.

Chapter 6, *LAN Internetworking*, is concerned with the networking elements that LAN internetworks employ. Some, such as bridges, hubs, and repeaters have evolved from intra-LAN applications. Others, like routers and gateways, were specifically intended for internetworks. All are developing increased function and performance; this has led to a new hierarchy in WANs that allows the LAN to extend beyond its traditional boundaries. An extended LAN may encompass a city or a metropolis and have global connectivity via the public telephone network.

Chapter 7, *MANs to WANs*, is an account of the types of broadband networks presently being implemented and their impact on the existing public telephone network. One result of their rapid distribution is the vast (and sometimes bewildering) array of new carrier services that seamlessly internetwork LANs and WANs, satisfying the demand for sophisticated data, image, and video communications. Newly formed services include switched multimegabit data service (SMDS) offered by the telephone company/post, telephone, and telegraph authorities (Telcos/PTTs), and BISDN/ATM offered by LAN equipment vendors and Telcos/PTTs. All are dependent upon emerging technologies: SONET/SDH forms the basis for linking high-speed LANs, while ATM realizes the full potential of bandwidth-on-demand services. Together they integrate LAN and WAN network elements, using common equipment and making possible a rich variety of network architectures.

SECTION 3: BROADBAND NETWORK TECHNOLOGIES

Chapter 8, *Broadband Frame and Cell Technologies*, characterizes the broadband technologies that are impacting business productivity. One of them, fast packet, is already a success in private T1 networks. New broadband technologies such as frame and cell relay, SONET, and SDH (the synchronous digital hierarchy, SONET's European equivalent), as well as ATM, have entered the Telco/PTT networks. ATM, for example, allows the Telco/PTT to tailor a variety of services to the specific needs of business and residential customers. Now the same technologies, honed on the WAN, are returning to the LAN mainstream and are routing and switching information at unprecedented speeds.

Chapter 9, *Broadband Transport Technologies*, describes how the present asynchronous telephone network is becoming synchronous. Within the public telephone network there is a hierarchy of rates referred to as the asynchronous standard hierarchy. Although each of its frame-encapsulated signals is built on the preceding one in the hierarchy, at each level they are discretely separate, making it difficult to manipulate (i.e., multiplex and demultiplex) them. Nor can they be easily synchronized. As a result, the public network is synchronous only on a piecemeal basis and, therefore, lacks the management capability and bandwidth flexibility for many of the newly demanded services. In contrast, LANs provide inexpensive connectivity at least in a local area. But LANs, too, have come under pressure from the ever-increasing bandwidth demands of new applications. Because of its capacity to manage large amounts of bandwidth and its simplicity and cost-effectiveness, SONET will, in a few years, totally displace the existing installed base of nonstandard fiber and electronic equipment. On the other hand, since SONET is a transport technology, it will not necessarily displace emerging technologies such as frame relay, SMDS, FDDI, BISDN, and ATM which can and will be carried by the SONET network.

Chapter 10, *Synchronous Optical Networks*, continues to detail SONET/ SDH networks, concentrating on the new variety of network elements (NEs). Unique features such as direct multiplexing and grooming of DS0s, add/drop capabilities, and the standard optical interfaces will simplify the design and management of broadband networks. SONET equipment will form the tributaries, interoffice trunks, metropolitan and suburban backbones of public and private broadband networks. However, there remains a considerable investment in embedded asynchronous equipment which is still usable and which the telephone companies have been understandably slow to replace. The North American and European asynchronous transmission rates fall below those of their SONET counterparts. Although SONET NEs support the major existing framing formats, the reverse is not true. It is not possible to cram SONET signals into the existing network. It will be pressure from private networks that will act as a catalyst for SONET deployment and within the confines of corporations, the technology will inevitably take hold.

SECTION 4: STRATEGIC CONSIDERATIONS

Chapter 11, *Planning Considerations*, discusses how to arrange for the optimal use of broadband networks (LAN and WAN) along with the economies and efficiencies that they afford. The fundamental changes taking place in telecommunications with regard to technology and business orientation are creating new business realities. No business can any longer afford to be an island; all are interconnected by a vast global communications network. How effectively the network is used will be determined by how conscious network and information management employees are of their new roles. This awareness may very well determine a given enterprise's competitive posture and success.

Chapter 12, *Digital Convergence*, traces the ways in which telephone companies and cable television operators are competing for supremacy on the broadband digital highway local loop. Who will dominate the local loop? What will the converged networks look like? What will the services be? Will convergence occur naturally or by means of takeovers of cable TV businesses by larger telecommunications companies? This chapter attempts to answer these and the other questions that are motivated by the digital convergence of cable TV, entertainment, and telephony. Not too surprisingly, the answers may well depend as much on political and economic issues as on technical ones.

The explanations within this book are intended to be clear and concise. Unnecessary detail is avoided so that the readers can concentrate on what they believe is important. To make this a truly interactive experience, you are welcome to E-mail your comments to me at RPDINC@aol.com.

WHY BUSINESSES ARE MOVING TO BROADBAND

The Business Case for Broadband Technology

VITALIZING THE BUSINESS VIA DIGITAL COMMUNICATIONS

From remote manufacturing to posting electronic advertisements to collecting payments, businesses are determinedly "going electronic," relying upon their communications networks to differentiate them from their competitors. What began with demands from technically inclined users is rapidly moving into the mainstream. The goal of the business community appears to be nothing less than the complete integration of a mobile, global work force into the enterprise. Information managers, besieged by vendors and pressured by end users, realize that change is inevitable but often do not know where to turn. Sometimes a company has a migration plan, but too often the changes are haphazardly executed by a disparate army of consultants and vendors rather than with coordinated in-house expertise. Too few firms have taken the time to actually detail the skills and information (both paper and human-based) that make up their pool of knowledge. As a result, data that has taken years to amass is often irretrievably lost in the corporate maze.

Think about your own company. What is your situation? Does information flow freely to wherever it is needed, rather than up and down a hierarchy? Can customers dial directly into your computers to verify the status of orders or do they have to listen to music while they wait for a customer service representative to answer their call? Is your company electronically linked to suppliers in a just-in-time inventory network? Can your people be taught remotely by the best instructors and can this learning be augmented with computer-based training? Are your employees able to fill out their expense forms on PCs and get them approved and filed electronically? Will the legal department accept a contract that was sent by E-mail? If the answer to any of these questions is "no," then this book should be for you. Here you will find out about the communications products, technologies, and services that have become increasingly necessary for survival in this very competitive world.

Within these pages is a broad range of information about the LANs and WANs, routers and multiplexers that form the basic elements of today's worldwide web of communication networks. You will find clarification of the jumble of jargon that describes the digital technology advances that are able to reliably move huge amounts of information (e.g., ATM, ISDN, TCP/IP, SONET, etc.). Although the future of corporate computing lies in networking and even though company networks are pervasive, you will see why the benefits can at first seem to be elusive. This encompassing change is not only a matter of technology, but extends deeply into the corporate processes.

REENGINEERING VIA INFORMATION TECHNOLOGY

Companies are coming to understand that they must reengineer themselves because they cannot run twenty-first century businesses with 1980s style communications networks and organizations. Reengineering provides a way to restructure an enterprise by examining its work flow, deciding what is significant, and then streamlining the processes for maximum efficiency. The main task is to identify an organization's primary customers and define those customers' needs. Often, the work approach needs to be moved from a reactionary mode that waits for problems to arise to a proactive position that anticipates and prevents breakdowns or failures before they occur. To accomplish this, a network that facilitates cross-functional lines is vital.

One advantage to moving the work flow onto networks is that companies then can incrementally move from ad hoc scenarios to more complex, structured processes without completely changing the way their people do their jobs. Ideally, by networking business processes and automating routine tasks, users should become empowered to perform higher quality work, thereby maximizing their contributions. Because the interplay between user needs and applications defines the network (Figure 1.1), there are many reengineering considerations to be taken into account.

- Users: For businesses which are already keeping sales records on a computerized system, a terrestrial network may not always allow their sales force, which may be highly mobile, to be consistently able to access information on demand. In contrast, a wireless network could provide the mobility they need.
- Applications: The multimedia applications such as video conferencing that organizations may benefit from investing in, require large amounts of network bandwidth.
- Migration: Many organizations use point-to-point digital leased lines to

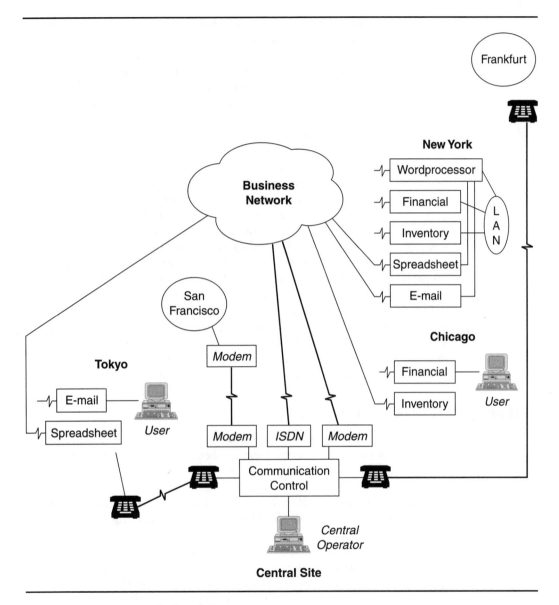

FIGURE 1.1   Networked work flow.

provide connectivity between facilities, but newer technologies, such as frame relay, deliver data faster and more economically. Often, as networks that are based on older technologies become more difficult and expensive to maintain and even, over time, obsolete, it makes sense to migrate to advanced technologies that perform better at lower cost.

- Growth: Growth may force network changes (e.g., the multiple networks that must be integrated into a larger homogeneous network when one company is acquired by another).

Network reengineering cannot be accomplished in an ivory tower, but should be executed with the specific goal of making the business more effective. It must be handled coherently and consistently throughout the organization because networks have become extremely sophisticated and now pervade all aspects of the business. Most importantly, a successful project requires a commitment at every level. The reengineering, itself, will call for a team that includes people who have the skills to develop, implement, and support the immediate network upgrade as well as have insight into what the ongoing support needs will be. For smaller efforts, only one individual may be necessary, but for most projects the team should be comprised of a project manager who manages and directs the enhancement, a customer contact to represent the interests of the network users, engineers to develop solutions, installation crews to install and test the equipment, and an operations representative to effectively integrate all of their efforts into existing network support mechanisms.

NEW INFORMATION INFRASTRUCTURE

The advance of digital communications technology is having a dramatic impact on businesses, their workers, and their suppliers and trading partners. Its force is being felt in many areas:

- The Organization—Old corporate barriers are breaking down, allowing critical information to be shared instantly across functional departments of product groups.
- Operations—Manufacturers are shrinking cycle times, reducing defects, and cutting waste. At the same time, service departments are streamlining ordering and communications with suppliers and customers.
- Staffing—Companies are using inexpensive computing to create virtual offices that link workers in far-flung locations.
- New Products—Just-in-time availability of information is collapsing development cycles. Companies are electronically feeding customer knowledge and marketing comments to globally dispersed product development teams so that they can rejuvenate product lines, target customer priorities, and take complete advantage of these resources.

- Customer Service—Customer service representatives are tapping into company databases in order to solve customer demands in real time.

The improvement of computer network performance presents a seemingly endless array of challenges. The ultimate criteria will likely be the degree to which the enhancements satisfy the demands of daily business applications. People want their information in formats that they can readily understand and that is available whenever they need it. To meet these requirements, there are several technologies that are already coming to the forefront in business.

- Multimedia—To the end user, multimedia is not separate from the computer. Instructions can be given with the touch of a few keys, the click of a mouse, or even voice commands. Information can be rapidly accessed from integral CD-ROMs as 3-D graphics, sound, or video. In terms of trying to move information in the form of multimedia across a network (e.g., videoconference across continents), the performance will depend on technologies such as asynchronous transfer mode (ATM), synchronous optical network (SONET) and its European equivalent synchronous data hierarchy (SDH), frame relay, and so forth that provide the network with sufficient bandwidth to transport the data without the receiving party having to wait beyond reason.
- Groupware—Groupware is actually a catchall term that encompasses products that range from E-mail systems through scheduling programs to electronic bulletin board software. The most sophisticated groupware packages have advanced features for automating the flow of work, routing documents, or allowing users at remote locations to collaborate on single tasks at the same time as though they were right next to each other.
- Middleware—In the past, applications such as billing, would run on a single mainframe computer that used software supplied by a single company. Today, networked applications must often rely on software from multiple suppliers whose products may not be compatible. Middleware gives these heterogeneous mixes the appearance of harmony by translating between different computers and software applications. In this way, corporate computer networks can provide easy access to a vast variety of information and electronic services—both within and across corporate boundaries, as corporate computing evolves from mainframe centric to a client/server mode.

UPDATING OBSOLETE COMPUTER NETWORKS

In the 1990s, companies moved away from large mainframe computers to networks of personal computers using distributed resource technologies such as client/server computing. Although mainframes are still in use, continuing to perform heavy computational jobs and provide a repository for large amounts of data, the ways by which they communicate with other network elements have changed significantly. They are now compatible with *local area networks* (LANs) and have open communication protocols as opposed to their past proprietary ones.

LANs initially were used within single buildings to connect computers and peripheral devices in order to share information. Now people use LANs to send *electronic mail* (E-mail) around the world, to access files stored in remote servers, and to collaborate on files using a special type of software known as groupware. In the past, to exchange information with computers outside the building, the voice switch or *private branch exchange* (PBX) funneled calls and data to local and long-distance services. It also managed all internal voice traffic as well as company voice mail systems. A LAN was connected to the PBX via a gateway computer and modems. But this was a relatively low-speed connection and not suited for the large bandwidth requirements of today's multimedia information.

Currently, LAN speeds have been ratcheted up to the point that they require the use of high bandwidth fiber links when connecting into the public telephone networks. For example, within a building, a local area network (LAN) transfers data at broadband speeds of 10 to 100 Mb/s. Then the data enters the local and long-distance public telephone networks and the transmission speed drops to 2 megabits per second. Modern fiber-optic cables now provide higher speed links in and out of the local telephone company central offices where calls are routed between local subscribers and long-distance networks. This has created a myriad of higher-speed communication options for corporate networks such as private lines and T-carrier, ISDN, and packet frame level services.

Updating existing networks is not simply a matter of technology but also depends strongly on attitude. Sometimes it is the corporate culture itself that inhibits implementation. Quite often, difficulties can be due not to equipment, but to the people who have been involved with the changeovers. Engineers may resent having to use standardized software. Managers may feel insecure and express disbelief in the viability of network upgrade projects. Although a full range of telecommunications products and services exists, deriving the promised benefits from them depends on a complete education of and buy-in from all concerned.

Even when the will and the means to change are strong, there will be areas where a strong foundation in broadband and its associated and component technologies is critical. In the area of digital communications there can easily be a confusion of alternatives for corporate and *information technology* (IT) communications people who want to support their business applications. In the past, when many companies updated their networks, the technology quickly became out-of-date or completely obsolescent. Companies do not necessarily need applications of the latest technology, but rather they require business solutions. For example, in the case of LANs that use switching equipment to segment the networks and move packets of data between users, in order to clear the congestion that will inevitably arise, planners need to know what high-speed switching devices to select, either Ethernet or ATM. This is a technology choice that will need to be made based on where the network is today and where it is going. Or, in another situation, telephone companies that are unable to resolve the problem of connecting networks between distant locations because their current switching gear is no longer appropriate, now have to choose between ATM, frame relay, SONET/SDH, and other prime technology options. There is no single solution. One must find the most appropriate ones by navigating through an apparent maze of vendor hype, immature technologies, and user demands. The wrong choices could require a "fork-lift" upgrade rather than a gradual migration. Ensuring that this does not happen to your business requires investing not only in upgrading equipment, but doing the same for your staff. It means training them so that their knowledge of digital communications is up-to-date and so that they can make informed choices.

ASKING FOR NEW TECHNOLOGY APPROPRIATIONS

Once these obstacles are overcome and you have a trained staff, you still must face other challenges. It is not uncommon for *information system* (IS) departments, as they are driven by the move from mainframe to client/server networks, object orientation, work-flow software, and hand-held computers, to have to cope with multiple operating architectures, complex LAN structures, and a mix of legacy and new equipment. Companies are being forced to reinvent their networks in almost the same ways that they had to when the IBM 360 and 370 architectures were originally brought out in the 1960s and 1970s. End users have become more computer sophisticated, thereby also contributing to the increase in demands on the IS departments.

Although some firms are now first investigating broadband networking and the associated investments, others have already taken the plunge and are implementing pilot projects. Their IS departments have to decide how to upgrade their networks to the varieties of modern broadband networks that have the bandwidth for multimedia business applications and for management to the desktops from a central console. Given the high cost of these systems, management will inevitably, and rightfully, question their worth to the company. (The cost for a large information network that is targeted for 10,000 or more employees would entail expenditures of hundreds of thousands of dollars for server software and thousands per workstation.) They will have to weigh buying all new equipment against keeping what they already have in place. To do this, in addition to calculating replacement costs, they will have to determine how much would be saved in maintenance and support costs by choosing to keep or buy or do both. The decision also depends on the existing mix of computer and communications equipment. For some cost areas, such as computer system up-time and maintenance, it is a simple paper exercise to determine the savings (Table 1.1).

TABLE 1.1   Sample Support Cost Considerations

Support Item	Cost $
Cost of ownership per user per year for legacy network	?
Cost of ownership per user per year for new network	?
Total cost savings per year per user	?
Support savings from new network	?
Cost of moving to new network per user	?
Time in which costs will be recouped	?

In other cases, the difficulty in calculating the price of change increases directly with the amount of outdated equipment that needs to be migrated. Moreover, the savings may not be directly apparent, but may be more difficult to quantify, such as improvements in customer service response time.

Implementation will vary depending on a firm's upgrade goals. An important consideration is sufficient bandwidth to handle multimedia applications (Table 1.2).

Multimedia information, as opposed to simple electronic communications such as E-mail, can be of significant benefit to a business in a multitude of ways. It can help to speed product development and minimize time to market. It can facilitate cost cutting, both in obvious ways, such as travel expense savings, and less obviously by expediting product development. And it can serve to increase presence and maintain face-to-face contact with partners, suppliers, and customers. The opportunities are everywhere. For example, fans of the Minnesota

TABLE 1.2   Bandwidth Requirements for Various Applications

Application	Megabits per Second
Text	0.02
Still Image	0.05
Telephone	0.6
CD-Quality Audio	1.4
Desktop Videoconferencing	41.5
High-Resolution Photo	60
Broadcast TV	221
HDTV	944

Twins baseball team are able to purchase tickets to games at 30 kiosks that the team has installed in local grocery stores. The kiosks connect directly to all of the team's accounting applications so that customers can complete the transaction without assistance. As an added benefit, the kiosks show videos of Twins games when they are not being used to order tickets.

WHY SHOULD SENIOR MANAGEMENT CARE?

When is it time to tell upper management that you need multimedia? Two criteria can help you determine whether your particular conditions are ripe.

1. Do you have users who are not technical, but who could make more productive use of information if it were available to them in a form that was more intuitive or natural?
2. Are you using information that is not in traditional alphanumeric characters or do you have to incorporate data information with alphanumeric information?

In other words, are you faced with situations where information could be used more effectively if it were provided in an easier-to-use context; or, if multimedia is already being used to some extent but not on a wide scale, is it because the installed network cannot accommodate multimedia's demands? If either situation is true, you most probably have an opportunity to save money and/or time by engineering multimedia into your system. If both are true, you are probably behind the competition.

Questions to be answered are what effect would these systems have on an organization's competitive position? What is the strategic fit with the company's short- and long-term plans? A company may need to promise a return on the cost of a new product within a year in order to justify it at the middle management level. To the senior management, the promise of less tangible benefits may be more effective. Cost might not be as much of a factor if it could be shown that the information provided by the network could lower the risk of introducing a new product or that it could provide the ability to examine and analyze market trends in real time, thereby allowing the company to substantially save money.

In this and the past decade, leading corporations have used networks to support novel business approaches that dramatically changed the competitive landscape. Communications technology created winners and losers as we saw Wal-Mart rise and Sears fall, Microsoft triumph and IBM slump. And yet, with the assistance of emerging technologies, these very losses could well be reversed. Today, "getting wired" in order to speed up internal process and electronically reach both within as well as outside the organization is imperative.

Electronic Commerce

With the growth in awareness that communication networks can remove the limits of time and distance, commerce over the network is becoming more and more real for the business community. Increasingly, they are using *wide-area networks* (WANs) to engage in electronic interchanges with suppliers and customers. For some, the Internet, a worldwide network of computers linked by telephone lines, provides an inexpensive avenue for commercial dealings; it is quickly becoming a widely used network for services and products. It is changing the way business is done in industries that include, among others, shopping, publishing, advertising, and broadcasting. To take advantage of the Internet's promise, companies that are intent on conducting transactions over the network must establish a presence on the *World Wide Web* (WWW), where multimedia information may be posted and advertised to reach a potential audience of 25 million customers. However, it should be kept in mind that until electronic commerce becomes more secure, Web sites will remain largely informational and promotional.

Telecommuting is another fast-growing trend, with companies using groupware to connect telecommuters to corporate computer networks over cellular networks and digital phone lines. In many businesses, these networks enable companies to constrain costs by allowing them to vacate expensive office space and provide the means for their electronically interconnected employees to work either from home, on the road, or wherever is most suitable. Thus far, telecommuting has gained the most ground in sales and other travel-intensive jobs, but it is likely to expand further with the introduction of improved and cost-effective desktop videoconferencing.

In the commercial sector, the electronic capability to work almost anywhere has given rise to the rapid growth of home-based businesses. Market research reveals that about 38 percent of U.S. households (about 37 million) have at least one person working at home. To serve this market, local phone companies are engaging in extensive efforts to make high bandwidth services such as ISDN available at reasonable rates since more and more people are equipping extra rooms in their residences with computers, fax machines and modems.

Increasing Productivity

The current direction of business communication networks is away from mainframes and toward the interconnection of distributed LANs, servers, PCs, and workstations. This tendency has been reinforced by numerous success stories. For example, in one computer company, Sun Microsystems, faster access to order data and inventory cut revenue processing time in half, generating $700 million in cash annually. In another industry, British Petroleum, an oil company, electronically processes almost 40 percent of the 440,000 invoices it receives annually, not only cutting down employee time and paperwork, but also realizing added savings by negotiating bulk rates and eliminating duplicate purchases. In another instance, L. M. Ericsson, a telephone equipment provider, employs 17,000 engineers in 40 research centers which are located in 20 countries around the world, all linked into one network. Its development teams in Australia and England can work together on the same design and then dash off the final blueprint to a factory in China. Finally, there is a food supplier, Delmonte Corp., that has begun to electronically receive daily inventory reports from the grocers it serves. When merchandise falls to a predetermined minimum level, the network immediately issues a restocking order. Retailers now need to maintain only one and one-half weeks of inventory in comparison with the four weeks worth they used to have to warehouse in order to avoid shortages. This has resulted in savings of over $625,000 per year for each food retailer.

CONCLUSION

Information is, and will remain, king; but, while focusing on its justifiable importance, many businesses fail to appreciate that, in this decade, the powers behind the throne are the broadband networks that distribute it. Many firms are pondering the benefits of upgrading their networks to support new applications. They face decisions with regard to both the nature of the applications and the networks that support them. Decisions need to be made whether to replace current applications, rewrite or rehost them, or implement middleware for cross-platform access to enterprise data. The network decisions are even more complex. Perhaps, there is value in the existing network. Its past usefulness, the IS staff's familiarity with it, even the embedded investment in it, may contribute to making it worthwhile to only overlay it with a faster platform while retaining the core functions. On the other hand, the solution could be to distribute the intelligence throughout the network by interconnecting LANs and WANs to funnel multimedia information from client to server. Conversely, new and more demanding applications might make it more cost effective to simply bring an all new broadband data network on-line.

It is still too early to be crystal clear as to which broadband network standards, equipment, and technologies will be accepted, so knowledgeable management-level individuals and system engineers must work together to plan and execute the migration. Although there are no simple guidelines, the business applications themselves will ultimately determine the direction of the network migration. For this reason, subsequent sections in this book provide a step-by-step road map for delivering a network that meets your company's needs.

CHAPTER
2

New Applications

INTRODUCTION

The coming of age of global communications networks has removed geographic distance as a barrier to efficient contact between people, businesses, and nations. Advances in lasers and fiber optics have led to very high-speed networks that can transmit the contents of a complete encyclopedia in the blink of an eye. In this era of worldwide, multienterprise networking, the ability to employ such technology within the public telephone network has become crucial for modern commerce. The electronic transmission of facsimile, E-mail, and *electronic data interchange* (EDI), among other forms of electronic information, has altered the very nature of trade. In today's knowledge-based industries, the rate at which individuals and organizations acquire information is and will continue to remain a significant advantage.

The growth of private corporate networks has slowed down and been replaced by an interest in more efficient interconnections with the public network. However, the existing asynchronous public network infrastructure lacks the bandwidth and flexibility that is being demanded by newly developed applications. These programs combine software-based technologies that simplify the interface between the individual and the information and vastly increase productivity, but at the cost of consuming large amounts of bandwidth. In addition to these bandwidth-hungry applications, customers increasingly want desktop-to-desktop transmission of an assortment of information, including voice, data, image, and video. Because all of this requires a coalescing of public and private networks and interfaces, public networks must make huge monetary investments in migrating to newer, more powerful infrastructures capable of supporting the bandwidth and the processing demands of multimedia software. The objective of this migration will be to create a global network so powerful that it will, in effect, transform the way we do business. It will empower both the corporate mainframe and the "road warrior" laptop, allowing them to connect not only within an organization, but also to remote branches, trading partners, and customers. Unlimited by place, time, equipment, or media, they will send and receive telephone calls and mail over terrestrial fiber, satellite, and radio waves completely unaware of the route that the information travels.

This unified, worldwide network will meet information needs of any scale, configuration, and application. On command, it will distribute communications between different types of computers and local networks. From the palmtop to the mainframe, from New York to Tokyo, this joining of people and resources will encourage faster and better decision-making, improving profitability and competitiveness. For the first time in history, information will be within the reach of all people who need it.

This future is not a dream, but the result of a natural evolution based on the driving technologies of fiber optics and silicon integrated circuits. Already, the use of fiber optics has lowered network noise levels by orders of magnitude, allowing a multitude of new streamlined communications protocols. And the ubiquitous silicon integrated circuit continues to double in density every nine months. The complexities of the systems that can be constructed in silicon are now powering the communications industry in the same fashion that they once powered the computer industry. The same integrated circuit revolution that made computers more affordable is now fueling broadband LANs and WANs, providing network elements that operate at ultrahigh data rates.

Although integrated broadband services will be the driving force in the next generation of networking, broadband transport is not a recent technology. It has been deployed for decades in the form of public asynchronous backbone networks and private LANs. What is new is the confluence of technologies that allow widespread broadband transmissions over the telephone network. The underlying structure of the public network has changed, with fiber-optic cable replacing copper lines. Fiber-optic transmission between switching exchanges has become the rule in the industrialized world; its efficiencies, performance, and economies have created a multibillion dollar equipment market that is growing at double-digit rates. Fiber routes are less noisy and can transport information without the time delays incurred with in-route, error-checking protocols. Businesses now have at their disposal unprecedented computer power at reasonable prices, increasingly available fiber media, faster LANs, and advances in silicon integrated circuits. These, in turn, have allowed the practical implementation of new protocols, interfaces, and switches that, themselves, are based on emerging broadband technologies such as SONET, SDH, and ATM. Taken together, all of these enabling technologies are allowing the WAN to become an extended-distance LAN.

The need for this infrastructure is based on the traffic generated by everyday industry. Corporate network managers must prepare by understanding the trade-offs between, and the motivations for, broadband networks. The largest hurdles are often not technological ones, but rather the will and ability of companies to retrain their employees, to retarget development projects, and to pry funds loose to improve the network infrastructure.

NETWORK TRENDS

In the past, WANs that were capable of transporting information at T1 or lower rates (less than 1.544 Mb/s) provided adequate transport. The networks were dominated by voice or point-to-point data and packet-switched facilities that connected dispersed business locations and computer equipment. Their architectures supported the access of remote terminals by mainframe or minicomputers. Today, the host-centric networks of the 1980s have given way to distributed computing environments. High-capacity token ring and Ethernet LANs now deliver data between desktop workstations and personal computers. Originally, the interconnection of these LANs was limited to building-wide or campus networks, but as the scope of communications became global, LAN internets began to deliver the data to the worldwide communications network, exposing weaknesses in the existing broadband network infrastructure. WAN bottlenecks were created, causing outages, overloads, and intolerable delays for real-time application, and a glaring lack of bandwidth management.

Today, businesses that previously were concerned with their mainframe computer networks are discovering that the WAN is as much a part of their day-to-day activities as the LAN. It is difficult for them to ignore the changing telecommunications environment—an environment fraught with terminology, protocols, and economics that are so very different from the more familiar LAN—and remain competitive. The telecommunications industry is undergoing a change as profound as that imposed upon the mainframe computer by the personal computer. With digital traffic growing at an astounding rate and exceeding the capacity of the existing public telephone network, broadband technology is clearly the key to the future of public and private networks around the world.

BROADBAND APPLICATIONS

If computing needs were to remain static, the narrowband WAN infrastructure and, consequently, the disparity between LANs and WANs, would probably continue for years to come. But with the emergence of bandwidth-hungry applications in everyday business activity, change is inevitable. For example, in 1995 in Europe alone, investment in intelligent buildings was $15 billion because of the ability to save up to 30 percent of building operating costs. This environment optimally integrates people, property, and technology by managing a variety of computer services on a local network. These islands of computer power then connect into the worldwide public telephone network. For the user, the manner in which information is accessed and the burden it places on the network remains transparent.

With the proliferation of such applications, there is a parallel increase in bandwidth demands upon the public network. Regardless of the size of the communication links, increases in the number of connections will more and more tax network capacity. Even LANs, with their relatively large megabit buses, are not immune to bandwidth overload. Ethernet, token ring, and FDDI all share media—a situation that is always prone to creating bottlenecks. A shared LAN can bog down under the burden of large numbers of users and of information-intensive transfers. With so much information being shipped over a single route, the ability to manage and dynamically allocate it on-the-fly is critical. Therefore, individual users can be better served by having exclusive use of the bandwidth, rather than by sharing it. That is the basis for some of the new switching technologies and the driving force for the new broadband transport technologies that can manage huge amounts of bandwidth.

Broadband networks have the capacity to channel the daily flood of information from local and campus environments and distribute it to remote desktops and information processing resources (Figure 2.1). As the amount of data builds up, it enters broadband channels employing communications speeds beyond T1 rates (1.544 Mb/s). The resulting transmissions may either remain within a campus or building, be dispersed through a metropolis, or span the globe. The source of information is not important. What matters is the capacity and routing capability of the networks. Broadband networks with throughput approaching the speed of light are already changing the rules of how business is done. In a world economy where real-time and managed access to massive amounts of information is important, these networks have become vital information superhighways.

The immediate requirements for broadband networks stem from the day-to-day use of an expanding variety of office software programs. The backlog of seemingly simple applications is producing larger and larger data fields. To remain competitive, leading-edge executives need access to the most current information without geographical or time constraints. They also now rely more and more on graphs, pictures, tables, and even video with text. As a result, the size of the files used to store this kind information has increased tenfold. This growth has created massive operating systems for personal computers. While the original DOS loaded in two diskettes, Windows and OS/2 uses more. The commonplace word processor alone has become exceedingly powerful, incorporating capabilities once reserved for spreadsheets, databases, and desktop publishers. Whereas just a few years ago a large spreadsheet was 300 kilobytes, today's financial spreadsheet files contain dozens of graphs that are all updated in real-time and consume megabytes of memory. Databases and

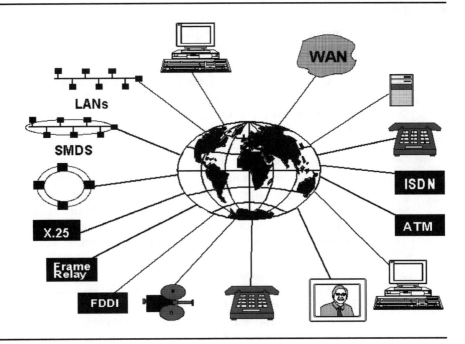

FIGURE 2.1   Global broadband networks service voice and data.

desktop publishers use up even larger amounts of memory, as do programs in multimedia formats.

Transmitting files that are generated by these programs can be a lengthy process unless enough bandwidth is available for use. Bandwidth determines how quickly information may be transmitted from one point to another in a network. A computer file that is sent over a 2,400 bandwidth line takes roughly one hour per megabyte of information to be transmitted. Sending megabytes of data even faster than that demands networks with the capability to supply and route even larger bandwidths. There are also monetary issues to be considered since charges are levied by telephone companies for the amounts of bandwidth consumed.

The balance between available bandwidth, business applications, and incurred costs has moved most large corporations to use not one but several networks, each with a distinct purpose: one for voice, another for SNA traffic, and possibly another for LAN traffic. Telco/PTT network architects face the daunting task of consolidating these diverse networks into a single broadband fabric. With success, their reward will be an increase in data traffic beyond today's growth rate of 20 percent a year.

Workgroups

With the fundamental alteration of global communications networks in terms of the ways in which information is transmitted and the form in which the transfer occurs, businesses are able to ask: What is my competitor doing in the Japanese market? Where is the sales group's videoconference? Can I see the signature of the person who approved the purchase order? These real-time questions require real-time answers; the responses emanating from facsimile machines in Vietnam, from cellular telephones in Argentina, from satellite dishes in Russia, and from video terminals in Los Angeles are all part of a communications revolution that is upending conventional rules for where businesses are located and the ways in which they operate. Today, with a worldwide public communications infrastructure of high-capacity fiber cables, digital switches, and satellites both small local firms and giant multinational corporations can conduct business with comparable degrees of productivity and efficiency. In these times of reengineered and flattened organizations, competitive businesses require the total democratization of information, reflecting actual work patterns rather than traditionally defined ones.

The timely transfer and management of information relies on the coalescing of two worlds that have hitherto been separate and distinct—computing and telecommunications. This ongoing convergence is creating a complex and changing environment for corporate communications. In order to meet the challenge of connecting disconnected users, businesses must look at software applications and the networks that support them in fresh ways.

Software

Workgroup software, sometimes referred to as *cooperative software* or *groupware*, allows a single action to automatically generate a host of other actions. It takes advantage of new, more powerful operating systems to make the physical topology of the network transparent to the user, who no longer needs to know where resources are on the network or how the communication occurs. With this structure, all applications become workgroup applications regardless of most temporal and spatial constraints.

Workgroup software boosts productivity by allowing an organization to leverage highly skilled people. For example, Lotus Notes is based on an object-storage mechanism that uses replication technology to update remote copies of E-mail databases housed on a Notes server. It is also used for forms and images. People on the network have access to an electronic bulletin board over which they can automatically send and receive information in whatever form they desire, to and from as many other users as they want. Coworkers can share information (Figure 2.2). Marketing and sales groups can share a common database that tracks and monitors all contacts with a single customer. Accountants can share financial and tax information. Secretaries can exchange letters.

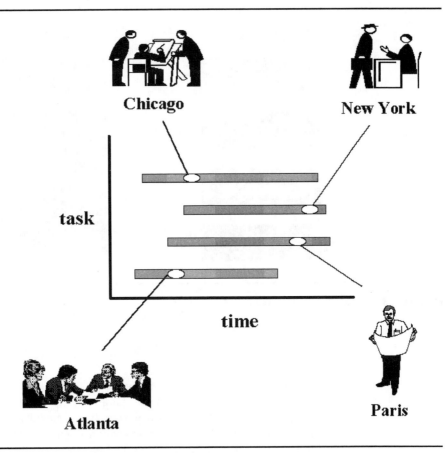

Chicago

New York

task

time

Atlanta

Paris

FIGURE 2.2   Workgroup interaction via the network.

Engineers can exchange design files. A single voucher on a computer can be simultaneously forwarded to a supervisor for approval and to a secretary for filing. When signed by the supervisor, it can automatically be sent to accounting where a voucher can be generated. In another instance, a team project can be coordinated by an electronic schedule that automatically updates each member.

All of this results in better and faster decision-making, but with the trade-off of a subsequent increase in the volume of routine business transactions that traverse the network. Consequently, although it increases productivity, such software does consume large amounts of bandwidth between end-stations because the multimedia applications are generated without user awareness (Figure 2.3).

On a social level, at the same time that workgroup programs improve a company's productivity, some of their benefits can also raise concerns among a

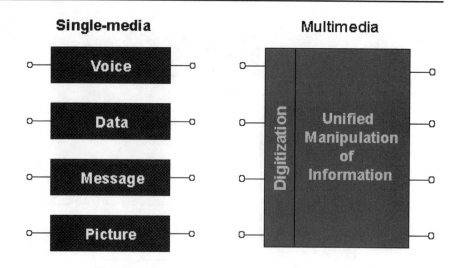

FIGURE 2.3   Multimedia formation.

management force that is not prepared for the inevitable results of shared information (Table 2.1).

Nevertheless, it is historically interesting to note that similar concerns were raised by MIS managers when personal computers were initially introduced. Whenever the existing paradigm is perceived to be in jeopardy, people are wary. This is mentioned here only because it is necessary to ensure that when there are plans for an organization to be electronically reengineered, all of the personnel and social considerations are taken into account and realigned accordingly.

TABLE 2.1   Workgroup Issues

Benefits	Concerns
Gives end users broader access to data	Could let workers know more than would be preferred
Flattens organizations, speeding decision-making	Could threaten authority of chain of management
Improves individuals' ability to collaborate	Could result in loss of individuals' recognition and competitive edge

Communications

The appropriateness of the *public-switched telephone network* (PSTN) for voice conversations is derived from its physical connection capability by means of a relatively simple interface and without the user having to understand the details of the connection. A call can be set up to any location by dialing a telephone number and, barring language differences, the parties can then speak directly to each other. A considerably more complex situation exists for computer conversations. First, the call setup requires data synchronization, that is, communication and error-checking protocols that are not necessary for voice connections. Furthermore, if the computer systems employ different proprietary protocols, there is a connectivity problem; just as people need to speak the same language in order to understand each other, so do computers. The detail of ensuring that a data conversation will be heard and understood often requires the support of trained network engineers. Since most people are not sufficiently technically proficient, there is a definite need for a common tool that allows communications across diverse computer platforms and makes it simple to develop and use applications. This is the goal of an emerging communications technology—collaborative computing. Many businesses are willing to invest in modern communications networks and appliances, but only the most electronically sophisticated realize the impact of behind-the-scenes collaborative computing on the efficiency of communications. To transport crucial messages to the user as needed requires desktop-to-desktop connectivity based on an information appliance (e.g., workstation, personal, or hand-held computer or telephone), software, and a network—either data or voice. These are the building blocks of the collaborative workplace.

Applications

Computing in the past focused on individual business applications, such as payroll and accounting. Interaction between these applications was limited because the only intelligent user information appliance available, the personal computer, employed microprocessors with thousands of transistors that provided relatively small amounts of memory for program storage and processing. With advances in silicon integrated circuits, microprocessors with millions of transistors were developed, allowing the running of more complex software programs with expanded intuitive user interfaces. Today's personal computers store hundreds of millions of bytes, supporting entire business processes that share applications including product development, account management, and customer service. With such large amounts of memory and more powerful microprocessors, network users are able to run on-line business processes that generate background applications which, in turn, receive updated information by means of complex database queries.

TABLE 2.2   PC Screen Draw Time (Minutes)

Screen	Resolution	Colors	Transmission Bandwidth		
			(Bits/sec)		
	(Bits)	(Bits)	1,200	9,600	28,800
VGA	640 x 480	8	34	5.3	2.2
Super VGA	800 x 600	16	106.7	13.3	4.4
GUI Acceleration	1,024 x 768	32	350	43.7	14.6

Modern business processes employ applications with *graphical user interfaces* (GUI). In concert with multimedia data types such as images, digital voice, text, and data, GUIs accelerate both the development and acquisition of information. While GUIs improve productivity because they allow people to understand concepts more rapidly, they demand ever-larger amounts of network resources. Common activities such as receiving a color picture can take an inordinate amount of time—over five minutes for a 256-bit color image using a modem transmission bandwidth of 9,600 bits/sec (Table 2.2). While various image-compression techniques can reduce the transmission time, the only lasting solution is more network bandwidth.

Telecommuting, E-mail, and Facsimile

Although telecommuting has long been touted as the solution to traffic jams and the energy crisis, it has not been promoted as a way to increase productivity. And yet, Arthur D. Little, the Boston-based management consulting firm, estimates that teleworking could save the U.S. economy $23 billion per year. The inevitability of telecommuting is being aided by the political process. Today, 108 million Americans commute to work by car, an increase of more than 30 percent since 1980. The Clean Air Act of 1990 was implemented to try to reduce car commuting by 25 percent in the most polluted U.S. cities. The alternatives are to car pool or to work at home one or more days a week; the latter is finding growing favor as it seems to fit comfortably into our changing work and life styles in general. Not so coincidentally, as highway traffic lessens, network traffic will increase, for telecommuting will contribute to data build-up in the public network—a situation that the swelling acceptance of messaging technologies such as E-mail and fax will exacerbate.

On the plus side, the economies and efficiencies of electronic messaging are becoming more widely appreciated, making geographically dispersed businesses easier to operate and manage. Progressive companies are already using electronic messaging not only to keep in touch with their telecommuting work force but also to stay in contact with remotely located employees as well as with customers and vendors. Using messaging to augment and complement voice conversations not only improves partner relationships, but also reduces the time and cost of making decisions. For example, there is an actual company that has 28,000 corporate users that communicate with 12 million electronic mailboxes worldwide. For this company there was never any question about extending their successful corporate messaging system to the outside; the problem was how to implement it internally to serve their customers and suppliers better. Proprietary solutions were quickly discarded as it became apparent that only an international, public messaging network could keep pace with technology advances, as well as work with the company's existing hardware and allow the users of equipment from various vendors to communicate. The solution was to connect their internal system to a public messaging network that consolidated and managed their faxes and E-mail, thus ensuring that both internal and external communications were simpler, more effective, and future-proof.

Multimedia and Hypertext

In the future, people will interact with computers as easily as they do with each other. Information will flow freely, formatted for the rapid assimilation of ideas (Figure 2.4). The interface will be composed of today's voice, image, and text all melded into multimedia presentations. The decision to employ multimedia is no different from other business decisions. The selection of any new technology must involve addressing issues of purpose, cost, and resources. The real value becomes apparent when the technology fulfills a purpose in ways that are clearly superior to other means. Used appropriately, multimedia can introduce new avenues for workgroup communication and collaboration (e.g., sound can greatly benefit sight-impaired users).

Multimedia is based upon the same digital technology as computing. The information that resides in an analog format—video, audio, and text—is translated into binary ones and zeros, a language that is easier to send, store, and manipulate for computers. Multimedia relies upon the electronic integration of different media to enhance the ability of users to interact with the information in a format that suits their particular needs, via the use of specialized peripherals such as CD-ROM, videodisk players, or VCRs. Standards are still being worked out. Many of today's microcomputers will not support the ones that are finally established (Table 2.3). Network managers want to implement applications that support high-quality, full-motion video that requires broadband networks. Even with compression, VHS-quality video takes 1.2 Mb/s of bandwidth; advanced digital video can use 30 to 130 Mb/s. LANs like Ethernet, which were designed to carry data, have problems transporting full-motion video, which requires the delivery of packets in a particular order with small, consistent delays. Packet

Paper documents are linear structures

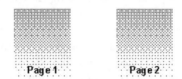

The content is often non-linear

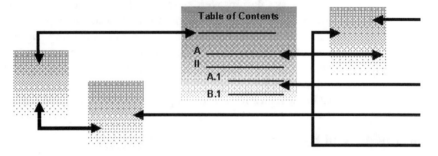

FIGURE 2.4   Non-linear information.

video on LANs needs faster, isochronous protocols to deliver the continuous data streams of voice, video, and data with little delay. Such isochronous protocols establish a virtual channel that behaves like a dedicated point-to-point circuit between devices. Among the technologies competing for multimedia traffic are ATM, fast Ethernet, and FDDI-II (detailed in Chapters 4 and 6).

Hypertext is a term for a retrieval system in which individual words are indexed and threads created to have one word lead to another. By merging text and graphics and not requiring specialized computers, hypertext constitutes a logical migration point to multimedia. This level of integration is highly useful in developing computerized training modules, for example. Hypertext is especially useful in conjunction with massive databases because it takes information retrieval to a new level by making it simpler to use and find. To accomplish this, many database vendors rely exclusively on Boolean searches to create indexes automatically that list the occurrence of every designated word or related sets of words in the database. While this capability is valuable, by itself it is not true hypertext. To create true hypertext, the contents of the database must be coded to link one subject with another.

Just as it took the installation of millions of LANs to create sufficient LAN traffic to impact the WAN, multimedia applications are now gathering the same kind of critical mass. Hastening this process is the introduction of multimedia

TABLE 2.3   Multimedia standards.

Standard	Proponent	What it does
Digital Video Interactive	Intel	Offers real-time video (RTV) and production level video (PLV) compression at a ratio of up to 160-to-1. RTV compresses video in real time, while PLV compresses stored video. RTV and PLV offer play back rates of 30 frames/second.
Quick Time	Apple Compter	Enables Macintosh users to create, store, transmit, and play back compressed video at rates between 15 and 30 frames/sec. Apple is transporting QuickTime to Microsoft Windows.
Video for Windows	Microsoft	Enables Windows users to create, store, transmit, and play back compressed video at rates between 15 and 30 frames/sec. Supports the Audio Visual Interleave compression standard.
MPEG	ISO's motion picture experts group	Several MPEG standards are being developed to format video clips at compression ratios from 50-to-1 up to 200-to-1.
JPEG	*CCITT and ISO's Joint Photographic Experts Group	JPEG has its roots in still-image compression and formatting and is now addressing full-motion video. The JPEG algorithm offers relatively high resolution and uses a compression ratio between 20-to-1 and 50-to-1.
Px64	CCITT	A suite of standards for video conferencing that includes the H.261 standard. Allows for connections between video conferencing codecs made by different vendors using single or multiple 64K bit/second circuits.

*CCITT, now ITU the Comite Consulatif Internationale Telegraphique et Telophonique, a United Nations International Telecommunications unit.

files that increase by 4 to 10 times in size each year due to the amount of charts, pictures, and even voice that are being incorporated into the text. Today, most of the personal computers and workstations sold in the United States have multimedia capability. Coupled with hypertext, use of that capability will inevitably increase network traffic. Every time such traffic enters the network, the demand for bandwidth will increase—as will the average network traffic load.

Even though, to date, the limited amount of bandwidth available in the public telephone network has tended to keep multimedia applications on the premise, the goal of advocates of multimedia is to allow networked computers to handle sound, image, and video as readily as they handle numbers and text. Consequently, a major factor for success will be the availability of sufficient bandwidth to manage large numbers of multimedia applications. Although the full impact of these applications is still several years away, it is imperative that the network infrastructure be in place well in advance.

Videoconferencing

Businesses spend hundreds of billions of dollars on travel, costs that videoconferencing could substantially reduce by transporting conversations and images instead of people. Video signals can be transmitted over private links (e.g., dedicated T1 and T3 links) and over the public telephone network. The difference between the two is the degree to which the transmission can approximate full-motion video. Video can be transmitted at lower bandwidths with signal compression such as that specified by the *motion pictures expert group* (MPEG). However, some forms of compressed video use image prediction algorithms that have trouble tracking rapid scene changes. Rather than the stop and start of slow-scan video, consumers favor transmissions that resemble what they have become accustomed to seeing on television screens. In contrast, with a broadband network-based videoconferencing system, broadband switches can be employed that use multiple fixed cameras. Viewers then can select which person they want on screen and the system will select the best angle and the right camera for the picture. The increased consumer acceptance of motion video transported over broadband networks could jump-start the videoconferencing market and outweigh the increased cost for the necessarily larger amounts of bandwidth.

Image Processing

Researchers estimate that 95 percent of the world's information resides on paper or microfilm. The task of moving it from these dated technologies to electronic form is accomplished by image processing systems. Image processing converts existing documents of all kinds to digital representations which can be quickly viewed on a computer monitor, then printed and distributed in a variety of ways, either as maps, fingerprints, CAT scans, drawings, or anything else that can be represented in two dimensions. The components of a basic image processing system include scanners resembling fax machines that make a digital record of every small sector (called dots) of an existing document, a controlling computer (microcomputer, minicomputer, or mainframe), a monitor, and a means of storage. In most networks, the efficient use of storage media such as optical disks and jukeboxes is pivotal to a successful imaging system. The high gigabyte content of most CD-ROM platters is essential to archiving libraries of documents because a single, high-resolution image can consume tens of megabytes of storage.

Imaging is often viewed simply as a storage and retrieval vehicle, but document imaging systems can mix data, photographs, and full-motion video and can completely transform the way an organization operates. Evidence of this is the federal government who, since the mid-1980s, has been widely using image processing, replacing microfilm and microfiche which employ photographic technology. Since the management of financial records and contracts is the top application for federal agencies, imaging technology has reduced the storage space needed for documents by a factor of 12 and allowed quicker retrieval. For the IRS alone, the potential exists to save $41 million a year in the storage and retrieval of paper documents.

Given today's technology, it remains a sizable challenge to shuttle hundreds of megabytes of images across the enterprise network. The digital storage requirements of a single full-color image can exceed 16 megabytes, providing the detail for the rich variations of color and texture in fine artwork. Experts in the United States could study great works of art in Russia without leaving their terminals. The transfer of such large files over a public or private network requires bandwidths that provide a tolerable amount of delay or transmission time—that is, the hours or seconds required to complete the file transfer. Higher bandwidths cost more, but allow information to be transported with less delay. In the case of specialists, such as art assessors or medical radiologists, their time alone may be worth the increased network costs. For other businesses, the increased cost of bandwidth is justified by considerable overall savings in other operations. For worldwide catalog sales, instead of producing costly monthly paper catalog updates in several languages, central image storehouses can be accessed by users in different countries across high-speed, real-time links.

One of the world's largest pizza restaurant chains has managed to cut response time to suppliers from 5 days to 10 minutes, while receiving a 16 percent after-tax return on investment. A network of imaging equipment tracks this company's fixed assets and handles the more than 10,000 faxed invoices it receives from worldwide suppliers each month. As invoices come into the accounts payable department, they are scanned into the imaging system. Over 40 workstations are connected by fiber optic cabling to servers which are interconnected by means of an Ethernet LAN and can, thereby, access two mainframes. On the mainframe computers reside databases that store the locations of the chain's restaurants and can identify more than 90,000 vendors.

Optical Computers

Introduced in the 1950s, silicon integrated circuits have been instrumental in transporting much of the world from the Industrial Revolution into the Information Age. Many believe that the next great leap will come from the marriage of light and electricity in the optical computer, a technology already under investigation. A prototype developed at the University of Colorado consists of lasers, electronic switches, and optical fibers arranged in layers and tightly packed into an area about the size of a desk. As in electronic computers, information is represented by binary ones and zeros. The optical computer uses the presence or absence of light pulses to represent the binary pattern. The prototype optical computer operates at a clock rate of 50 Mb/s, controlling a 16-bit microprocessor. Data-encoded light pulses are stored in some three miles of spooled fiber cable. Each bit of information is carried in a 12-foot-long light pulse, which traverses the memory spool every 20 millionth of a second. The pulses are synchronized by the unvarying speed of light. From such prototypes a number of information age applications may emerge, including

- A high-speed graphics processor that uses millions of optical switches interconnected in free space by mirrors rather than fiber cable. This type of computer can link a virtually unlimited number of optical switches—at least in theory.
- A 20 Gb/s optical computer on a single silicon wafer.

Mobile Computing

The union of the communications, computer, and consumer electronics industries is rapidly creating products that allow people to transfer data without wires. Wireless communications are made possible by the emerging electronic superhighway and by the services, products, and information that will travel along it. The growth in wireless data transport is driven by the proliferation of portable computers, LANs, client/server computing, spread spectrum, and cellular telephony and by the falling cost of transporting data via broadband networks. This blend of personal communications technologies is creating the mobile professional—an employee untethered by place or time.

Wireless data communication is freeing workers from traditional constraints. Mobile applications such as the officeless desktop, one-way and two-way messaging, long-distance file uploads, and the ubiquitous E-mail allow them to reach anyone no matter how far away. The benefits of wireless E-mail are numerous: It can always find individuals; business deals can be negotiated while in transit; and people can stay in touch with the home office and with clients by way of messages that can be initiated and responded to as needed no matter where the sender or recipient might be. Wireless E-mail may be the most significant tool for the professional on the move, replacing pagers and cellular telephones, not to mention airport and hotel room searches for RJ-11 jacks!

Home Video

Although local cable companies offer limited versions of home video, only the more comprehensive switched telephone network can offer universal video access. Despite the desirability of this service, fiber-optic connections to the home are not being made as fast as many experts originally predicted because of the planning and study time incurred by *local exchange carriers* (LECs). For them, the stakes are high: Although there are potentially 140 million access lines in the United States alone, they could cost $1,000 each to convert.

However, this market is expected to expand rapidly. Government regulators have allowed the LECs to enter the home video market and, in return, the local loop has been opened to competition from alternate access providers, cable companies, and long distance as well as local telephone companies. A steady stream of projects is under way with the goal of merging PC and TV technologies to create a super information appliance. Fiber-optic vendors, cable TV operators, and telephone companies are teaming up in an effort to be among the first to install the fiber-optic cable that will carry a variety of services to the home market.

Digital TV broadcasts will enable intelligent video dial-tone systems. Such systems go beyond today's pay-per-view, a system that requires a user to call the local cable company to sign up for a movie that the cable company has scheduled for broadcast. With video dial-tone a viewer uses a remote control to program the TV to capture and download a movie from a menu of movies. When the movie ends, the viewer then can download other information from a shopping network, a stock market quotations source, or any other available service.

The race is on to build an electronic superhighway capable of delivering hundreds of new TV channels, as well as data and phone services. Once the loop is opened, and low-capacity systems are upgraded to support broadband transmission, the path will be cleared for electronic encyclopedias, shopping catalogs, travel, banking services, and more, all brought to the home on a real-time basis. Full-range audio and high-resolution video-on-demand will make home entertainment an entirely new experience. At the same time, video to the home will also open the way to truly effective telecommuting, favorably impacting the performances of the more than 30 million home-based businesses in the United States.

Voice Processing

Voice processing encompasses a number of separate but related technologies, including voice messaging, voice response, interactive/transactional voice response, text-to-speech, voice recognition, and synthesis. Most voice automation systems use digital technology to store, retrieve, and manipulate voice signals. Voice mail has traditionally been the leading application of voice processing. As a solitary application it is yielding to multiple applications. Many systems now also have auto attendant, audiotext, *interactive voice response* (IVR), and other applications simultaneously available. Of all the applications for voice processing technology, IVR systems are the fastest growing market segment and, after voice mail, the second largest market sector. The IVR market currently is growing at an annual rate of 25 percent whereas the voice mail annual growth rate has leveled off to 16 percent.

The IVR caller typically keys information in on a touch-tone telephone. With transactional systems, a more sophisticated form of IVR, users can input information to a host database. For example, a caller who uses voice response to check on his or her customer file and locates an error in the database can leave a voice message detailing the discrepancy without having to call the company again. Integrated systems also offer advantages for internal users: In a single telephone call, traveling sales representatives calling their own phones back at the office can receive an order previously left on their voice mail and check the company's database to determine the availability of the desired item. One of the most exciting applications is the combination of IVR with the personal computer for forms processing (Figure 2.5).

The integration of telecommunications and PCs is growing. In spite of the fact that a voice processing system is essentially a specially equipped computer system, telephone companies have not integrated voice mail effectively with computers because telephones are treated like commodities in most businesses. Nonetheless, future desktop systems will combine the telephone and micro-

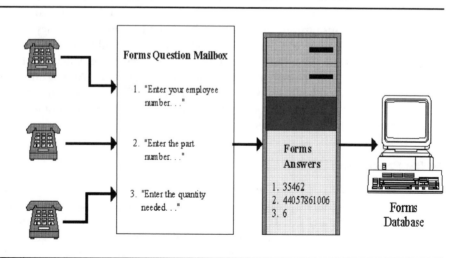

FIGURE 2.5   Automated forms processing.

computer into a tightly integrated desktop unit. At some point, the advent of portable personal communicators will provide hand-held voice mail capability. Users will be able to make telephone calls through the PC by clicking on an icon to place, receive, and put calls on hold or play them back.

Hardware and Software

PC-based systems are equipped with a processor, multiple port I/O cards, hard disk drives, a monitor, a keyboard, and voice processing software. The system connects to a Centrex, key systems, or PBXs through a single line extension or PBX interface. PC-based systems provide:

- A database for voice storage
- Network interfaces (modem, CSU/DSU) that allow the caller's telephone to function as an input/output (I/O) node on a network
- Menu-driven application generators that enable users to customize their voice processing applications

The fundamental hardware component is the voice card. The voice card plugs into an empty expansion slot in an off-the-shelf PC (Figure 2.6). The card

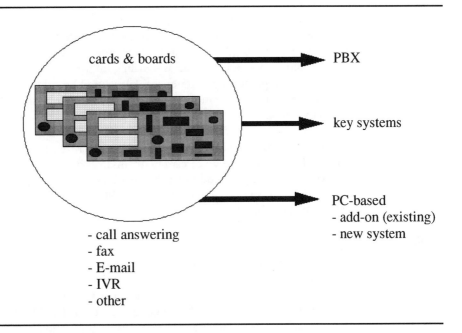

cards & boards → PBX

→ key systems

→ PC-based
- add-on (existing)
- new system

- call answering
- fax
- E-mail
- IVR
- other

FIGURE 2.6   Voice card applications.

TABLE 2.4   Voice processing system.

Hardware	
computer	ground start interface
loop start interface	T1 interface
direct inward dialing	speech recognition card
fax card	station adapters
audio multiplexers	audio couplers
audio interfaces	

Software	
operating system	speech analyzers
development system	device drivers
speech library	compilers
debugger	call progress programs
diagnostic software	voice prompt editors

converts voice and other analog signals into a digital format and connects to the telephone network. The card also connects to the internal computer bus as well as its own voice bus. (A voice bus is independent of the computer controlling the card and allows audio as well as signaling information to be passed between different voice processing components.) It is used by other plug-in cards that provide fax, voice recognition, and text-to-speech conversion. Add-on products allow human voice enhancements for E-mail, word processing, and spreadsheet analysis. The products capture short voice recordings and attach them to a message or document for later replay to emphasize or expand on text and images. The PC and the voice card by themselves do not form a voice processing system, but require additional hardware and software (Table 2.4). Voice messaging software runs under such popular PC operating systems as Windows, MS-DOS, and UNIX.

Applications

With voice messaging, subscribers simply dial into the messaging service to pick up or drop off information. Received messages can be automatically distributed according to customer schedules. At the same time that voice mail has been combined with other forms of messaging (e.g., E-mail and EDI), voice and fax integration have been picking up steam. While application development is still in its infancy, two general areas have emerged:

- Integrated messaging, which merges fax, voice, and E-mail in a single subscriber mailbox. Upon interrogating the mailbox, the subscriber is informed that a mix of messages is waiting.
- Information dissemination which, in most small systems, still consists of audiotext, although IVR usage is beginning to grow.

Integration

PC-based products can be interfaced to a key system, a PBX system, another PC, or to a LAN or WAN. Although today's LANs do not carry voice traffic, they can be used to combine PCs to increase processing power. In the near term, it is anticipated that LAN bandwidths will increase tenfold, allowing voice as well as data traffic to be transported. Scientific advances in photonics and silicon have contributed to a new age that is characterized by desktop to LAN-to-WAN communications. *Asynchronous Transfer Mode* (ATM) switches that allow concurrent routing of voice with data are already available and work equally well with LANs or WANs. (ATM is a powerful switching technology that is promising to undo the hitherto inviolate separation of data and voice traffic.)

SOCIAL IMPACT

Broadband networks are faster and more intelligent than the LANs and WANs they replace. Comparatively speaking, their improvements over earlier networks are as sweeping as those of the fax over the earlier telex. It cannot be repeated too frequently that the ability to manage information and provide bandwidth on demand liberates users from time constraints, improves relationships within organizations, and decreases the costs of gathering and dispersing information. Properly put to use, the delivery of information in all its forms—text, image, and video—promises an unprecedented high level of educational, health, and entertainment services on a worldwide scale. The explosive growth in the number of home-based businesses, estimated in 1995 to be 35 million in the United States alone, is a direct result of their being able to compete via the global telecommunications network. Home-based workers are not the only ones making innovative use of the telecommunications network. Traditional businesses have discovered that the savings in time and labor are well worth the investment in electronic messaging. They are finding that freedom from geographic and time constraints is crucial for competing in global markets.

This view also extends to nations. The European community intends to strengthen the interoperability of existing networks and, at the same time, reinforce its economic and social aims. The Maastricht Treaty, signed in 1991 by members of the *Commission of European Communities* (CEC), proclaimed that the community shall "contribute to the establishment of trans-European networks in the areas of transport, telecommunications and energy infrastructures." This infrastructure will provide a common dissemination point of knowledge, culture, and history of the European people.

Less-developed countries, too, have the opportunity to leapfrog into broadband communications—terrestrial fiber, synchronous transport, scaleable protocols—gaining benefits perhaps *before* more developed countries that have a considerable investment in their existing asynchronous network infrastructures. Developing countries, for example, could end up with telecommunications networks that are superior to those in Europe or North America, providing them with a framework for economic competition.

CONCLUSION

Traffic on the public telephone company network is increasing because of new workplace applications: Databases, word processors, and spreadsheets that were once reserved for specific individuals or could only be used by one person at a time are now within the reach of groups. The way people work, sell, and gather information continues to evolve, resulting in increased output and more efficient communication with both customers and workers. Via global data networks information is communicated to individuals, machines, businesses, and nations.

Today's computer and telephone eventually will fuse and evolve into a completely portable and self-contained workgroup communications appliance, proficient enough to access a vast variety of personal communications services. This small hand-held console of the future will instantaneously link a person to any print or visual media in the world over the public telephone network. The basic components of this new way to communicate knowledge are already being activated—the downsizing of the personal computer, collaborative computing, miniature video displays, and synchronous optical networks. With advanced integrated circuits, digital electronics, and data streams, the real-time packaging and transmittal of information over broadband networks will generate unbelievable gains in business productivity.

3

New Computing Networks

INTRODUCTION

Once limited to voice and low-speed data transport, the global communications network is adjusting to the accelerated introduction of data. In the past decade, data has supplanted voice as the predominant source of public network traffic. Hospitals transmit X-ray images to specialists hundreds of miles away, students tap into distant research libraries, and executives log onto office networks from home. However, they are still the exception rather than the rule. For this trend to become even more widespread, a new global network infrastructure is needed in which, for the first time, standardized protocols can flow over a single switching and multiplexing fabric, allowing the internetworking of all types of data communications equipment. All points in the network—public or private, LAN or WAN—will have to use common technology.

OPEN SYSTEMS INTERCONNECT (OSI)

The first step in implementing such networks is to establish a common model for networking. Earlier proprietary networks such as IBM's *system network architecture* (SNA) required equipment from a single vendor. While the network worked well, it did not allow the customer to use the latest technologies and equipment from other vendors. Nonetheless, IBM's SNA was used as the basis for modeling an open system, that is to say, a network formed by building blocks that adhere to standard protocols.

The foundation for open systems is the *open systems interconnect* (OSI) reference model (Table 3.1). This model of a communication system, developed by the *International Standards Organization* (ISO), is most valuable as a way to char-

TABLE 3.1   OSI reference model.

7	Application	Uses messages— Communications interface between users and applications, i.e., allows file copying, virtual terminals, etc.
6	Presentation	Transforms data between systems; also may provide decompression and de-encryption.
5	Session	*Uses remote procedure calls (RPCs)—* Manages dialog between systems; provides remote software functions on the remote system, i.e., no disk space, paper out, etc.
4	Transport	Uses *segments*— Ensures the reliable delivery of data, usually through connection makes and breaks, acknowledgment messages, sequence numbers, and flow control.
3	Network	Uses *datagrams*— Provides for data movement across different network segments, organizing the Data Link layer and often including the logical source and destination addresses.
2	Data Link	Uses *frames*— Organizes the Physical layer in logical groupings, often including the physical source and destination addresses; provides flow control and error detection (sometimes error correction).
1	Physical	Uses *bits*— Defines the mechanical and electrical requirements of the media and interface hardware.

acterize networks and as a goal for uniform network implementation. Today, all major networking players attempt to define their equipment using the OSI model. The model describes the seven steps or layers required for end-to-end communications. Each layer communicates with the layer directly above and below it and has its own communications mechanism. This structure allows error checking and recovery at the lowest possible layer, thereby improving throughput. Most communications systems do not use all of the layers; in fact, the network portion of end-to-end communications is found only in layers four and below. Higher layers such as application, presentation, and session are normally associated with the sending or receiving computer. Also associated with the layers is equipment that allows networks to be interconnected—gateways, routers, bridges, and repeaters. Throughout this book, the OSI reference model will be used to describe networks and the equipment that forms them.

ACCESSING THE BROADBAND NETWORK

Modern communications networks satisfy the desktop-to-desktop service demands of users. A network is formed from *customer premises equipment* (CPE) situated at customer locations, from transmission equipment that is part of the network, switching equipment that forms the network routing fabric, and management elements that provide provisioning and control. There are a variety of environments (Figure 3.1) for CPE. Although there are instances of a single personal computer or terminal connecting to the network, for broadband networks it is more likely that traffic will be generated from a LAN.

Over the past two decades, proprietary computer architectures and protocols have kept customers captive. Unable to select the best equipment for the task, they were forced to purchase from a single vendor. The rise of LANs has broken these chains because LANs allow computer equipment from different vendors to be interconnected. Connections to the LAN are achieved with interfaces or adapters that plug into each device on the network—usually PCs, workstations, printers, or servers—and allow them to communicate with other devices. While some LANs exist as completely separate networks, it is more likely that LAN traffic is pooled at an interface such as a bridge, router, or

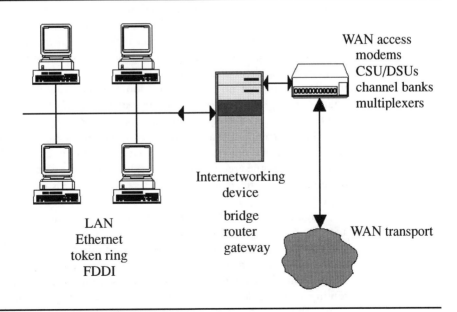

FIGURE 3.1   CPE networking components.

gateway to access devices on *other* LANs. These serve as the interface with the WAN if the device being accessed is remotely located. Since the transmission protocols employed by LANs are effective over relatively small distances, the interface must convert between the LAN and WAN transmission protocols. Furthermore, protocols employed by the computers being accessed may differ from those employed by transmitting devices. In the past, the numbers of different communications protocols hindered conversation between devices. One of the promises of broadband communications is the lessening of the differences as the protocols employed by LANs and WANs coalesce. Similarly, the promotion of open architectures and protocols in the broadband WAN promises compatible equipment and services.

High-speed Mainframe Interfaces

Although the hierarchical computer networks of the 1980s have grudgingly given way to LANs, the need for mainframe computers remains for those activities that require the manipulation of large amounts of information, that is to say, enormous databases or very high-speed computation applications. As mainframe computers have become more powerful, direct high-speed linkage to a mainframe port has become more important. Mainframe computers are accessed via internal channels that, like telephone lines, operate at various data rates. Channel rates of 64 Kb/s or 1.544 Mb/s (the DS0 or T1 telephone equivalent) are common, but for activities such as connecting to a LAN or supercomputing, for example, higher rates are desirable.

The *high-speed serial channel interface* (HSSI) is a physical interface and a de facto industry standard for serial transmission at rates reaching 52 Mb/s between data terminal and communications equipment. HSSI defines the physical layer for transmission but not the transmission protocol, thereby allowing compatibility with high-speed network services such as *switched multimegabit data service* (SMDS). Instead of using the interface to link routers to point-to-point T3 lines, HSSI provides a way to connect routers to meshed SMDS networks. In contrast, alternative ways for high-speed channel connection—Fibre Channel and Escon—define architectures for sending data over dedicated connections at distances of 500 meters to 10 kilometers and at speeds from 17 to 133 Mb/s, respectively.

High-speed channel interface technologies constitute a fusion of networking and channel approaches to mainframe system interconnection. They combine traditional peripheral connection, host-to-host internetworking, distributed processes, and multimedia applications into a single multiprotocol interface. The Fibre Channel framing protocol, for example, includes data-type qualifiers for routing frame payloads into specific interface buffers and link-level constructs associated with individual operations. Its network-oriented facilities include multiplexing of traffic between multiple destinations, peer-to-peer connectivity between any pair of Fibre Channel ports, and internetworking with other connection technologies.

PSTN Transport and Switching

Information transported over the public telephone network may be in analog or digital form. Analog transport is used for low-speed voice, while digital signals that are represented by binary ones and zeros are used for data as well as voice transport. There is no inherent superiority of one format over another. Although real-world information is usually in analog form, the digital format is preferred for high-speed transmission purposes because it is the natural environment for computing. Once information is converted to digital form, it can be manipulated and transported by digital machinery that takes advantage of the economies and efficiencies of silicon integrated circuits. It also can take advantage of the rapid improvements in digital circuitry and available fiber optic cabling which allow higher speed signals to be transported, while maintaining synchrony over the long-haul telephone network.

Access from a private network, such as a LAN, into a public network is done by means of dial-up and private lines. Most telephones dial-up the PSTN, while data terminals that require continuous connection may use transmission lines that are leased from the telephone company. Sometimes a pair of terminal devices are connected by means of a private line that is dedicated to them, such as the linking of a bank teller terminal in a remote branch to the mainframe computer at the central location; in the past it was more likely that the terminals were connected by means of the public switched network. Today, connection over dedicated private lines at data rates reaching T1 or above is becoming more common.

Switching systems employed in the LAN or WAN provide economical connectivity. Ideally, any terminal device connected to the network can communicate with any other, but large-scale interconnection often requires a hierarchy of switches. For example, the North American system employed by the PSTN uses a series of offices where the switches are housed (Figure 3.2). The class 5 office serves the local loop, which is the most common means for subscribers to interconnect with the public network. Here information from low-rate telephone lines is combined with high-usage trunks. These trunks form the broadband portion of the existing asynchronous public network and may achieve broadband rates of one gigabit per second.

Role of Traffic and Performance Analysis

In traditional telephone networks, the relatively low-speed lines that are employed make it important to analyze traffic in order to optimize network performance. Statistical units of measure such as Erlangs are used to determine ways to squeeze out available bits of bandwidth for voice conversations. Because the

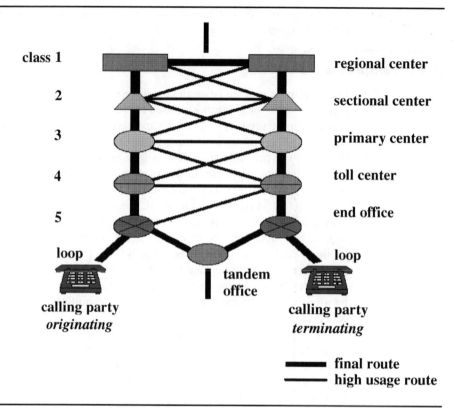

FIGURE 3.2   Switching hierarchy.

customer essentially pays for bandwidth, there is a cost savings when the same quality of conversation can be maintained at lower bandwidths. Knowledge of important parameters such as peak traffic times and *minutes of use* (MOU), among others, allow network architects to decide on how and when to provide new network routes and switches as well as how to alleviate undesirable network traits such as call blockage.

The same considerations also apply to data networks, but here the situation becomes more complex. For data calls, the circuits between the sender and receiver often are not maintained for the call duration, but rather established and disconnected as required to forward packets of information. Moreover, traffic may be automatically rerouted as paths become momentarily congested, making it more difficult to determine the routes that information takes when it travels through a large data network. Nonetheless, for low-speed data networks, traffic analysis has been refined almost to an art. Here, too, the ability to use all available bandwidth is important. With broadband networks, the significance of traffic analysis changes somewhat. At higher bandwidths, the issue is not just one of maximizing the use of available bandwidth and transport rates, but the management of such large bandwidths.

Broadband equipment can handle an incredible quantity of information. For the most part, the effective management of large amounts of information requires a tiered network architecture, beginning with equipment that has enough intelligence to be remotely controlled and extending to the host and public carrier management systems. Initially, network management was concerned with homogeneous element-to-element communication. As networks grew in complexity, concerns about element configuration, performance statistics, and alarms led to a class of manager that provided central control of remote network elements. Security, disaster recovery, and reliability are all modern network management considerations, particularly with regard to broadband networks. Broadband network devices have the intelligence to gather information about their own failure conditions, measure the performance of signals that pass through them, and assign bandwidth. They can determine not only their own failures, but also those of other network elements. The *operations support system* (OSS) that controls these elements must be able to interpret this information and take appropriate action. The Telco/PTT OSS, which in the past has not been involved with the customer premises, is now beginning to interact with the management systems employed by private networks. The result will be desktop-to-desktop management of data.

DISTRIBUTED COMPUTING NETWORKS

The number of shared applications resulting from the expansion of computer connectivity is growing exponentially, to the point that the bandwidth demands of these applications have already noticeably impacted private as well as public networks. In the past, WANs such as the public telephone network, based upon technologies capable of T1 or lower rates (e.g., less than 1.544 million bits per second), provided more than ample transport facilities for most business applications. Point-to-point and packet-switched facilities that connected dispersed business locations and computer equipment dominated WANs.

These WANs were typically used for accessing host computers from terminals and employed vendor-specific communication architectures such as IBM's SNA. Often the proprietary protocols associated with these architectures prevented the intermixing of equipment from different vendors, limiting the users' selection. Moreover, all communication was done by the host, which would poll remote terminals to alert them when a message was to be sent or received. This procedure not only was slow, but significantly limited the flow of information. In recent years, corporate computing has moved from the centralized mainframe toward distributed networks of people and equipment that share information. Most computing power sits on desktops in departments, as people perform their jobs using resources at different locations on different types of computers. Thus, open communication architectures and protocols such as the Internet's TCP/IP have become the rule rather than the exception. These protocols provide "connectionless" information transfer in the same way as the PSTN. That is to say, there is no physical connection between users until the call is set up, thereby saving network resources as well as speeding the communication process.

As was discussed in Chapter 2, the interconnection for LAN topologies was limited to individual building or campus networks originally, but as the scope of the enterprise expanded to global proportions, LAN internets began to deliver the data into the worldwide PSTN, exposing weakness in the existing infrastructure (Figure 3.3). Today, the traffic patterns encountered by a telephone company's voice switches are often more indicative of data transmission than voice. This pull from the data side is accelerating the adoption and deployment of new technologies on the PSTN at a much more rapid pace than in the past. For example, ATM, an ultra–high-speed voice and data switching fabric developed for the PSTN, has been put into use in LANs years before the standardiza-

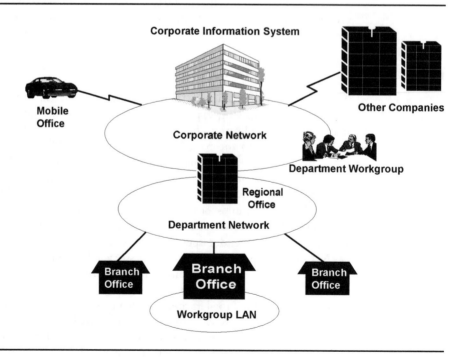

FIGURE 3.3   Collaborative computing network.

tion process will be completed for the WANs. Other new broadband network technologies are also unlocking the bandwidth in fiber optic cable—over 25,000 billion bits per second—used in public and private networks around the world. Frame relay, SONET/SDH, as well as ATM are allowing the WAN to become an extended-distance LAN. For businesses that were previously concerned with their mainframe computer networks, the WAN is becoming as much a part of their day-to-day activities as the LAN and is further building demand for new communications technologies.

CLIENT/SERVER COMPUTING

For the business community, survival may depend on the ability to reduce from days to hours, and from hours to seconds, the time it takes to gather and process information. A wide range of industries currently sends and receives information over globally scattered telephone networks that were originally designed for voice transmission. These businesses are pursuing more efficient ways to swap text and digitized images as they discover that data networking provides the kinds of productivity enhancements that they need. For them, productivity begins with the LAN.

LANs share computer equipment and information, helping reduce paperwork and enhance intra-office communication. Early LANs employed peer-to-peer networking between two equal devices. This was in stark contrast to the rigid network hierarchy used by mainframe and minicomputer computer networks such as IBM's SNA or DEC's DECnet. Although it freed individuals and small workgroups from unnecessary constraints, peer-to-peer networking was unwieldy for larger networks where security was a concern and more complex applications were the rule. In the 1980s, a more powerful LAN architecture evolved called *client/server computing*. Client/server networks include one main computer called the server. (This is the most common architecture. Nonetheless, in the client/server paradigm, there may exist several servers and several clients. In addition, a model has evolved referred to as client/server/server. In this model, the server function exists across several computing platforms. For example, a client/server application residing on a Novell file server may use *NetWare loadable modules* (NLMs) to communicate in the background to a DEC VAX database such as Oracle.) The desktop microcomputers are called clients. Ordinarily, all programs that the clients run are located on the server. The idea behind client/server computing is to separate the application from the service and then provide a simple path between the two. Such a concept, when put into practice, ensures that any service overhead is only incurred when the service is used. Since each interface in each service remains constant, new applications can be developed at any time. A client/server environment also allows services to be upgraded more easily.

There are many client/server applications: To use a factory floor as an illustration, a computer that supervises the work flow and a robotic arm controller can use different applications to draw only what each needs from a generic database server. In another situation, the joint use of resources could be among printing, E-mail, *computer-aided design* (CAD) services, and other functions. A client/server computing environment can consolidate a factory's *information system* (IS) to share hardware, software, and data, and ultimately to help reduce any redundancies. Client/server systems can also improve efficiency by distributing processing chores. This is most apparent in engineering departments. For example, one powerful shared workstation could be used as a CAD server to perform finite-element analysis—a computation-intensive task—while the engineers could work on a variety of other tasks with less powerful computers.

Bottlenecks

Despite the advantages of client/server computing, there can be drawbacks if attention is not paid to broadband network details. Often, more time and money is spent worrying about raw connectivity and how to migrate to client/server computing than whether the network will be fast enough to handle its traffic. Client/server computing requires faster networks for several reasons (Table 3.2):

- The distributed network increases the network loading and the physical layout can slow traffic as it passes from bridge to router to gateway.
- There is much more dialog between clients and servers than among machines in traditional host-to-terminal architectures. Important to the exchange is a two-phase commitment process which synchronizes and updates both the receiving and sending databases (necessary for cases such as the guaranteeing of a shipping order on both the sending and receiving ends of the transaction). Although the two-phase process ensures database accuracy, it adds significant traffic. In fact, network traffic can increase from 5 to 40 percent depending on the distributed database complexity.

TABLE 3.2   Client/Server network burden.

Activity	Net effect
two-phase commit and distributed network functions	up network traffic 5% to 40%
security checking	increases network overhead 10%; creates five- to 10-second delays
distributed physical layout (hodgepodge of software, hardware and connectivity devices)	hikes transaction volume 2% to 5%

- Sufficient bandwidth is also important for handling failures and peak loads. Distributed networks can use alternate routes during path failure, provided that there is bandwidth available. If the network fails and adequate spare bandwidth for rerouting the traffic is lacking, clients can continue to work on the distributed database, but the transactions will not be posted on the server database. Then, when the path is restored, updates will flood the network and cause congestion.
- Security checking between the client and server also slows transaction time and increases traffic. Clients and servers exchange information about users on the network to ensure that those who are at the client machines are legitimate users and not hackers.
- Backup and recovery are also a burden on the network. During backup, large bursts of data flood the network and result in a reduction of available bandwidth.

Unless these constraints are recognized and adequate bandwidth provided, the client/server can actually be a step backwards. One furniture manufacturer discovered that its client/server network slowed data traffic to such an extent that in six months of poor network performance the company lost nearly $2 million in business and at least $700,000 in employee productivity. This blind application of an advanced technology resulted in the early retirement of the responsible information systems manager. Since this is not an isolated incident, it is crucial that there be a complete understanding of network capacity and the impact that desktop intelligence has upon it.

Despite the drawbacks, organizations continue to migrate to client/server computing because of the need to network the ever-increasing desktop computer power and for workgroup collaboration.

Desktop Intelligence

The intelligence of desktop devices has increased dramatically because of improvements in the silicon integrated circuits such as microprocessors and memory. Over the past decade, the cost *per millions of instructions per second* (MIPS) for computers has declined at an annual rate of 25 percent. The cost per bit for *random access memory* (RAM) has been reduced 30 percent and that of magnetic memory used in disk drives 25 percent per year. As a result, mainframe costs have dropped 15 percent per year, minicomputers have fallen 25 percent per year, and personal computers 31 percent per year. Yet their power and speed have increased by orders of magnitude. Advances in silicon have made faster microprocessors possible, allowing the running of more complex software programs with larger output files. With successive generations of silicon microprocessor engines—Intel's 16-bit 286 and 386SX, 32-bit 386DX, 486 and 586—the desktop computer has taken on more of the characteristics of earlier mainframes, differing only in the number of input/output ports. The 586, called the *Pentium*, packs 3.1 million transistors in an area four times bigger than the 486. Each transistor is so tiny that it would take 500 of them to circle a human hair. The Pentium crunches numbers like a mainframe, executing 100 million instructions per second—five times the speed of the 486. Its successor, the P6, demonstrates roughly twice the performance of the Pentium. The room-sized mainframe of 50 years ago has evolved into a desktop box, 20,000 times more powerful and at a fraction of the cost (Table 3.3).

Today's desktop PC is nearly 20,000 times more powerful than the 30 ton ENIAC, the world's first computer. And the progress continues: The Pentium is 10 times as powerful as a high-powered model IBM sold in 1988 and is one-sixth the cost. The next generation P6 doubles the Pentium's speed for the same price.

The introduction of 100-MIPs computers on the desktop has already increased the load on the LAN. Soon the WAN will be significantly affected. Each computer MIP on the desk generates additional demand for bandwidth. Whereas 10 Mb/s Ethernet LANs are common, finding enterprises that use WAN bandwidth beyond T1 or 1.544 Mb/s is unusual. Obtainable network bandwidth has not kept up with new application demands and it will continue to lag until there is more widespread use of broadband technology. For LANs, these demands have led to a new generation that operates at 100 Mb/s and above. For WANs, they have led to investment in the synchronous transmission technology provided by SONET/SDH, and the faster switching fabrics provided by cell relay and ATM.

The interlocking relationship between more powerful computers and advanced software applications created the need for operating systems that make good use of the available computer power. A new generation of desktop operating systems—IBM's OS/2, Novell's UNIX, Microsoft's Windows '95 and

TABLE 3.3   PCs Pack More Power

	ENIAC	IBM System 370 Mod 168	IBM PS/2 Model 70	Dell 486P50	Pentium
Introduction	1946	1975	1988	1992	1994
Size	30 tons about the size of a boxcar	refrigerator size	desktop	desktop	desktop
Instructions processed per second	100,000	2 million	5 million	19.4 million	100 million
Retail Price	$3,200,000	$8,800,000	$13,840	$2,200	$2,100

Windows NT, and Taligent's object-oriented OS, among others—support the increasingly widespread utilization of desktop imaging, voice annotation, and multimedia. Early microprocessors used 640 kilobytes of RAM in the personal computers that employed them. Now personal computers support 4 to 32 megabytes of RAM. With these large amounts of memory and more powerful microprocessors, network users may now run on-line background applications that receive updated information in response to complex database queries. Computer companies have reengineered their applications to support distributed client/server computing, which increases bandwidth demand even if application requirements were to remain constant. But the application bandwidths have also increased. Many applications employ multimedia, incorporating audio, images, and full-motion video as well as text. Even voice overlays consume valuable bandwidth: As a rule of thumb, four minutes of digitized voice takes four megabytes of storage.

Collaborative Computing

A powerful benefit of client/server computing can be seen in situations where the creation and updating of the innumerable information sources or databases within internetworks of LANs and WANs need to be accomplished without either the user or originator having to understand the computer processes that underlie their delivery. Users want the capability of tapping into important sources of information without necessarily knowing the names or locations of the people or machines with which they wish to communicate (as opposed to standard electronic mail systems that are address-specific). This transparent accessibility to vital information is another one of the reasons for the rapidly growing popularity of a derivative of client/server computing—*collaborative computing.*

Collaborative computing is specifically designed to allow people to work cooperatively over a network. Collaborative computing takes file-sharing one step further than client/server does. In comparison with file-sharing which is often a process where users use and respond to documents off-line, collaborative computing opens up the world of real-time, on-line sharing of information and work. Once limited to a department or building, collaborative computing has spread as workgroup applications have become successful.

For the millions of potential information sources on the public and private networks, collaborative computing provides the real-time linkages to create, secure, update, and maintain them. The impact upon business productivity is dramatic: The layering of collaborative computing frameworks over existing corporate networks allows teams of people to blend their skills, knowledge, and work processes by managing the unseen network operations that enhance intragroup communications and performance. People can gather and present ideas in the most usable format as well as access, organize, track, and share them.

Examples of the productivity enhancements derived from collaborative computing include the following:

- Speeding up design: Developers in different companies can work together and share project management and product reports. This allows simultaneous joint development and testing while the participants share a single database that contains design and quality information.
- Improving Customer Service: Support and other types of information can be instantaneously and automatically distributed to customers on a worldwide basis, off-loading the burden of information distribution, while providing a higher level of service.
- Stimulating Sales: Communication with today's far-flung sales organizations is supercharged by shared access to pricing, competition, and service information. Sales people can work as effectively on the road or at home as in the office.

Despite its success to date, collaborative computing is still in its infancy. One of the current obstacles to its successful implementation is the narrow range of bandwidth currently available on the PSTN. However, each year bandwidth-intensive applications such as spreadsheet files and other business-related programs that contain charts, pictures, and even digital voice are increasing exponentially in size, driving the migration toward new broadband networks. This will directly benefit collaborative computing-based services since a major success factor will be the availability of sufficient bandwidth to handle large numbers of multimedia transmissions.

Collaborative software products range from traditional programs that have been enhanced with network-enabled features, to sophisticated packages designed to automate the office environment. Some traditional word processors include graphics and spreadsheet programs that have been augmented with workgroup features, including ones that let several users simultaneously update particular ranges over a network. In this environment, changes to a spreadsheet can be monitored, the date of the change tracked, and the person who made the change and the reason for the change identified. Among other applications are products designed to automate the way people work. Increasingly popular are group scheduling features, both in standalone applications that are designed to work over networks and in add-ons to E-mail features. Collaborative software applications can be tightly integrated with E-mail. For example, a user can send a schedule file while simultaneously using a word processor to compose a letter.

Moving beyond simple communications are products designed to facilitate such activities as routing information within an organization. Lotus Notes® is a good example. It combines traditional E-mail features with a forms designer and with application development tools built around a distributed, replicated database—replication for enterprise data sharing including Notes documents; various multimedia data types; and data from production databases. Bolstered by a wide array of add-ons, Notes provides the infrastructure for easily tapping into almost every advanced network and software technology such as fax servicing, the smart routing of faxes to the desktop, imaging, video to the desktop, and telephony, all tightly coupled with computing for customer service applications. Using public network services such as AT&T's Network Notes Service, the user can integrate voice mail, E-mail, and facsimile for global communications. For Internet users, Lotus Notes significantly improves the management and scalability of a company's World Wide Web presence. As a result, collaborative applications are dramatically growing at over 150 percent each year. Their popularity stems from the productivity improvements enjoyed by empowered users who interact with the information and each other in ways that suit their individual and corporate needs.

THE INTERNET

Formed from thousands of interconnected data networks, this decades-old network now links over 20 million users. Originally developed for government and university researchers, it is increasingly used for commercial purposes. The Internet, which connects people over data networks so that they can communicate with each other via computers, offers an inexpensive, fast, and effective way to exchange information with others, including customers, suppliers, and business partners.

The Internet originated in the U.S. government's *Defense Advanced Research Projects Agency* (DARPA), the research and development arm of the *Department of Defense* (DoD). In 1969, the DoD started building a packet-switched network called ARPANET to support various computer science and military research projects. In 1983 all ARPANET users were switched over to the open network protocols, known collectively as TCP/IP. The Internet's longevity is based on its robust design; it works even if sections of the network are damaged and allows the addition and removal of new users with minimal impact. It employs a multilevel hierarchical networking structure in which a backbone network links hosts and LANs so that local users can network easily and privately without affecting the larger networking environment (Figure 3.4). Moreover, the Internet supports just about every type of computer. These connect by either the IP Internet that provides direct, immediate, and interactive communications or by means of the E-mail Internet that provides limited service and indirect store and forward communications.

Once connected, Internet users may access a variety of key electronic services including the following:

Post office—to send and receive mail

Library—for browsing and retrieving free information

Bookstore—for purchasing information

Radio station—to receive broadcasts

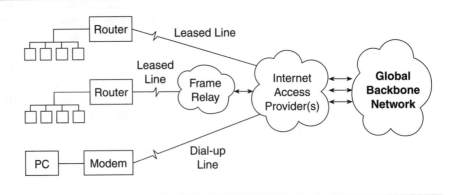

FIGURE 3.4   Internet access structure.

       Publisher—to publish information

       School—to create a classroom through connectivity

       Telephone company—for real-time conversations

The lure of virtually free information has proven irresistible. Millions of people enjoy *surfing* the Internet (i.e., browsing over its World Wide Web and gathering multimedia images and video as well as text). Until 1994, the Web was an obscure corner of cyberspace, little known even to most users of the Internet, just as the Internet, itself, was unknown to the computing public. Since then, the growth of this global network has been remarkable; its backbone has doubled in size each year, while its access networks have almost tripled. Over 160,000 new Internet users are added each month and companies ranging from giant corporations to small businesses jockey to win a share of the millions of households and organizations that are continuously being connected (Table 3.4).

    Inasmuch as unrestrained growth requires more and more bandwidth, it is inevitable that there are questions as to whether the Internet can support the growing number of voice and video applications as well as the millions of new users. More users and more sophisticated applications including multimedia require increased bandwidth; using a 14,400-bits-per-second modem, a modest size video clip of 10 megabytes could take more than an hour to download. Already some parts of the Internet are stretching at the seams. In 1994, traffic on the National Science Foundation's NSFnet grew 15 percent a month, an annual increase of 500 percent. The Sprint network portion alone experienced a monthly traffic jump from 5 to 30 terabytes per link for an overall yearly increase of 800 percent. To cope with this, content and access providers are compressing on-line documents, making them smaller and quicker to transfer. Nonetheless, the Internet and on-line services will not provide a viable distribution source for multimedia

TABLE 3.4   U.S. Internet Access Providers

Carriers	Access Providers	On-line Services	Computer Companies	Cable TV Companies
MCI	UUnet	Compuserv	IBM	TCI
PacBell	NETCOM	AOL		
Sprint	PSI	Microsoft		
AT&T	BBN Planet	Delphi		
Ameritech		Prodigy		

information until a broadband backbone infrastructure is deployed and the majority of consumers have an affordable and efficient way to download large amounts of information. When it comes of age, today's Internet will look like a dirt road in comparison with the true information superhighway that it will evolve into.

CONCLUSION

The LAN revolution became reality when Xerox's Ethernet became a de facto standard with the backing of DEC and Intel. DEC supplied the customer base and Intel supplied the integrated circuits that resulted in affordable connection devices. The result was a form of broadband network—the LAN. (For some, the term broadband network evokes memories of the early LAN implementations that used multiple frequencies to transmit data over coaxial cable. Even today, one of the largest applications of broadband networking is in the cable TV industry, where multiple channels are broadcast to millions of households.) Still, there is not one single reason for the accelerated demand for broadband services. Certainly the global competitive environment in which businesses must be well-informed and nimble to survive has contributed to it. Other factors include the growth of personal computers and the sophisticated software applications that make businesses more efficient and effective, but that at the same time require that pockets of information must be interconnected. These collaborative networks allow users to share knowledge in the forms of image, video, and text. They encourage organizations to automatically create network messages that track customers, create conferences, and access libraries of policies, documentation, and news. Never before has the networking of computers or the ability to send messages anywhere at will been so important to business success. And never before has the torrent of bandwidth demands deluged public and private networks so, pressing for intelligent and concurrently managed information superhighways and creating an explosive market for new broadband networks, equipment, and carrier services.

CHAPTER
4

New Communications Media

INTRODUCTION

Broadband markets have relied on the tremendous growth of desktop computers and fiber cable to bolster the demand for emerging equipment and services. In the past decade, the bandwidth of many business networks reached into the megabits. LANs went from one Mb/s ARCNET to 16 Mb/s token ring and are moving to even higher-speed 100 Mb/s FDDI and fast Ethernet. WANs increased from 64 kb/s to 1.544 Mb/s (T1) to 44.736 Mb/s (T3). Even higher-rate SONET/SDH connections became available. The dramatic demand for high bandwidth is due to the equally impressive increase in demand for personal computers, whose worldwide numbers exceed hundreds of millions. Both the network and computer usage have been stimulated by steep decreases in computer hardware costs because of integrated circuit economies and more sophisticated software applications. Soon, desktop computers with the processing power of hundreds of MIPS will be common, graphics resolution will increase twentyfold, video quality sixfold, and large file transfers such as CAD/CAM will grow a hundredfold. In addition, message size will increase fivefold, while actual network transit time will decrease by over 30 percent. The cumulative result will be an unprecedented growth in broadband markets.

One spin-off will be a market for broadband LAN equipment such as low-cost switching systems that operate at rates exceeding 155 Mb/s. Another will be for broadband services that will use a new Telco/PTT transmission and switching fabric (Table 4.1). But customers will need to understand their options for LANs and WANs and providers will need to understand the market dynamics. The customers who prosper will be the ones who solicit the combinations of features and services uniquely appropriate to their businesses. The providers who succeed will be the ones who offer these combinations.

TALBE 4.1   Data applications and Telco/PTT services.

Application	Traffic	Service
data internetworking	large amount of aggregate traffic, steady average flow	continuous bit rate—SONET, T1
LAN internets	distributed traffic bursts at high rates	SMDA, frame relay, ATM
multimedia, voice and video	delay intolerant, high rates over short duration	ATM real-time services such as BISDN

SUPPORTING TECHNOLOGY

Scientific advances in photonics and silicon have contributed to a new age that will be characterized by desktop supercomputers linked by networks that communicate at tens and even hundreds of *gigabits per second* (Gb/s). The virtual freedom from error and the high speed of fiber optics, together with powerful and economical desktop computers, have triggered a paradigm shift in which the WAN has become a homogeneous extension of the LAN. We are now entering the era of ultrahigh-speed, stripped-down communication protocols, and concurrent networking.

Fiber-Optic Cable

The bandwidth capacity of fiber is now being exploited for digital transmission. Thinner than a human hair yet stronger than steel, fiber has become the basic medium for long-haul and interoffice telecommunications networks. The special characteristics of optical fiber are its low impedance to lightwaves and its tremendous modulation capacity. It can support higher bandwidths over longer distances—enough to accommodate tens of thousands of voice channels. Researchers have transported information at rates exceeding 350 Gb/s. At that rate, the entire content of 1.2 million books could be transported around the globe in a minute! Yet this does not even challenge the potential information capacity of fiber cable which for wavelength division optical networks reaches 25,000 GHz. Unquestionably, one of the most dramatic changes in telecommunications over the last decade is the vast amount of fiber embedded in the public network which can provide almost unlimited bandwidth potential.

The fastest commercial fiber-based system today launches pulses down a fiber cable at roughly 2.5 billion pulses per second at much lower error rates (10^{-12}) than the error rates (10^{-8}) for the copper that it replaces. This permits high-speed, frame-based protocols that would not have worked over the error-prone copper lines. It also makes possible the simultaneous transmission of high-resolution video, digital audio, and data in two distinct network environments:

- In the WAN environment, fiber supports two types of high-capacity networks: asynchronous T3 at 45 Mb/s (European E3 is 34 Mb/s) and *synchronous optical network* (SONET in North America, SDH in Europe) with defined rates of 2.5 Gb/s, but capable of tens of gigabits per second.
- For LAN connectivity, fiber is being used as the means of extending traditional shared-media LANs via the *fiber distributed data interface* (FDDI) and *metropolitan area networks* (MANs) based on IEEE and Bellcore standards.

Its use in LANs and WANs has created a demand for two types of fiber-optic cable—single-mode and multimode. Single-mode fiber cable is used in the public WAN for two reasons: Information travels greater distances without requiring repeaters; and this type of cable removes the bandwidth limitations of multi-mode fiber (which operates below 155 Mb/s). On the other hand, multimode fiber cable is more likely to be found in buildings because it is simpler to install, lower in cost, and supported by major computer companies. Consequently, multimode fiber cable has become the favorite for fiber LANs such as FDDI.

Shared versus Concurrent Media Technologies

Historically, today's heterogeneous LAN and WAN networks evolved from a mix of networking technologies that were developed for completely different reasons. The public telephone network served primarily as a voice transport, while the LAN emerged in the early 1970s to handle data traffic. For this reason, LAN and WAN traffic and the technologies that support them have differed substantially. Technologies designed to accommodate bursty LAN traffic have been inappropriate for carrying real-time voice and video traffic which not only consume bandwidth, but also have a low tolerance for delay. However, the situation is changing as broadband applications develop and force all network topologies—local, metropolitan, and wide-area—to accommodate them. In the not-so-distant future, LAN traffic may be quite different. Instead of predominantly packet-based data traffic, LANs may carry significant amounts of real-time traffic generated by multimedia voice and video applications.

Shared-media technologies such as packet switching, Ethernet, token ring, and FDDI provide momentary access to a communication route and are appropriate for bursts of heavy communications traffic and periods of low activity. The route is used only as needed, allowing its cost to be spread among many users. Since this topology is susceptible to failure if a single station monopolizes the line, these networks use arbitration schemes for determining when stations can transmit.

Shared media can carry video traffic if there is no multiaccess arbitration between the workstation and the hub. In other words, video traffic can be supported if there is only one user on each segment. But this situation does not often occur in a shared environment. Consequently, shared-media LANs are not employed for latency-sensitive voice and video transfers. More appropriate transports include WAN isochronous networks which are synchronized to a real-time clock. Isochronous transport is a way to transmit such asynchronous information by synchronous means. Voice traffic is often transported by isochronous networks because conversations do not occur in regular, predictable intervals. One form of isochronous transport is the concurrent network.

Circuit switching and other concurrent technologies use dedicated routes to exploit the full-route bandwidth. The cost of the service is high because the user pays for the bandwidth even when it is not being used. Emerging switched services are not the solution for the user who needs continuous access for a major portion of the day, since they are priced higher than dedicated route service. Therefore, switched services are used primarily for route back-up. On the other hand, new broadband technologies combine the best features of shared and dedicated transmission media, making them useful for both LANs and WANs. ATM, for example, uses fixed-length cell-relay transmissions and buffering to ensure that multimedia applications are guaranteed a fixed response time, a requirement for transporting real-time voice and video. With ATM, the media is not shared. The media is switched, allowing each user full bandwidth as he or she uses the circuit. ATM also employs dynamically variable circuits through which sequential data packets can be routed on a point-to-point basis. This is in contrast with broadcasting traffic to every node on the network as with shared media or having a single application use the entire bandwidth, as with circuit switching. Users can almost "dial up" whatever capacity they need in real-time for the type of service desired.

Silicon Integration

High technology markets are fueled by the economies afforded by integrated circuits. Silicon architectures that duplicate simple atomic elements to form massively parallel systems are the basis for modern broadband technologies. With simple solutions in silicon, larger and more complex systems are economically possible. One result is the rapid evolution of the microprocessor, which is the engine of modern computers. These silicon chips continue to increase in function, power, and speed. The microprocessor has grown from the 4-bit bus of the 1970s through the 8- and 16-bit buses of the 1980s, to the 32-bit bus of the 1990s. Today microprocessors using 64-bit buses are on the near horizon, creating integrated circuits that possess the kind of processing power that up until only a few years ago was limited to mainframe computers.

Power is not the only significant factor. The price of computer components has dropped considerably. If these trends continue, it will be possible in the not-too-distant future to embed massive intelligence into everyday appliances. An interactive television costing only a few hundred dollars could incorporate a 32-bit microprocessor and 16 megabytes of memory. With a connection to a broadband network based upon SONET/SDH and ATM chip sets, the television could present a friendly graphical interface with full-motion video and voice response which could transparently search for programs. A student in New York City could learn Japanese with text from a data bank in Tokyo overlayed with pictures from an image bank in London—all in real-time.

WIRELESS COMMUNICATIONS

Wireless networks open up a new freedom of interaction between people and machines. They provide the natural medium for mobile voice and the personal portable computing devices that are growing in popularity. Wireless access to computer and voice communications networks is creating a ubiquitous information environment. Around the world, entire generations of technology development are being bypassed. When Hungary wanted to upgrade its aging telephone system, it licensed three cellular carriers to help jump-start the process. New networking concepts such as ad hoc networking, nomadic access, and mobile computing are entering the mainstream of communications.

Wireless transmission technology covers a wide market in networking, but it can be neatly broken down into two categories: radio and light. Radio wave transmission can be divided into licensed radio, which employs narrowband frequencies, and unlicensed radio, which includes frequencies in the 900 to 928 MHz and in the 2.4 to 2.483 GHz ranges. For light transmission, infrared comprises the main categories. The 900 to 928 MHz range is most commonly used for garage door openers, security gates, and cordless telephones. It uses spread-spectrum technology that enables network signals to operate in this range without interference. Spread-spectrum signals are sent over a range or spread of frequencies. There are two types of spread-spectrum transmission schemes: direct sequencing and frequency hopping. With direct sequencing, a single transmission is spread across a range of three or four subchannels. If the signal meets with interference at one frequency, it tries another. Transceivers recognize the way in which the spread is constructed and are therefore able to communicate. With frequency hopping, the range of frequencies is divided into many individual channels (up to 100), and transmissions are sent in a predetermined random sequence of channel shifts. Both the sender and receiver share the sequence of shift changes and no other device is able to access information unless the sequence pattern is known.

It is possible to operate several networks in the same physical area without interfering with one another. The primary forms of wireless networks are cellular radio, cordless telephony, and wireless data systems. Other categories include the *cellular radio packet data* (CDPD) standard and satellite.

Cellular Radio

Cellular is by far the most popular wireless service. There were over 19 million cellular subscribers in America in 1995 with over 14,000 new subscribers being added each day. By the year 2000, analysts estimate that the number of cellular subscribers will be between 45 and 66 million. In 1995, over 10 percent of cellular traffic was data; it is expected to eclipse voice traffic by the year 2000. The Yankee Group, based in Cambridge, Massachusetts, predicts that the revenues from wireless data service will reach $2.7 billion in 1998.

Cellular radio allows the subscriber to place and receive telephone calls, using bands of 824 to 849 MHz and 869 to 894 MHz in the United States. (Similar systems deployed overseas use frequency ranges between 200 MHz and 1 GHz.) The distinguishing feature of cellular systems compared to previous mobile radio systems is the use of many base stations with coverage of about 10 kilometers. Analog cellular systems such as the pervasive *advanced mobile phone system* (AMPS) use *frequency modulation* (FM) for speech transmission and *frequency shift keying* (FSK) for signaling. (AMPS was pioneered in the 1970s by Bell Laboratories. It has been available since 1983 and today it has more than 20 million subscribers.) Each call uses a different set of frequencies. In this way the spectrum is shared by employing a technology called *frequency division multiple access* (FDMA). In such cellular systems, continuous coverage is obtained by handing off a call from one base station to another.

Today's voice-oriented cellular networks were not designed with data transmission in mind. The noisy connections and momentary dropouts (as the transmitter moves from one cell to another or encounters physical obstructions such as bridges and tunnels) that cellular phone users routinely experience are only an inconvenience for voice calls. But they can be fatal to data calls. Manufacturers have tried to tame cellular networks with cellular modems that use advanced error-correcting protocols to try to compensate for signaling problems. While these products work, they are a patchwork solution at best. Connecting today's portable PCs and cellular phones often requires an awkward and costly interface device. Even then, getting connected can be difficult since not all cellular modems and on-line services support the same protocols: Some modems rely on MNP 10 or V.42bis, for example, while others use proprietary schemes. There are other drawbacks as well. For instance, once on-line, the speed and reliability of cellular's circuit-switched voice channels can vary. In addition, using cellular systems for data is quite expensive.

Digital Cellular

Digital cellular systems improve reliability by the use of low-rate digital speech-coding techniques and digital compression techniques. Virtually all forms of wireless communication, including cellular and paging services, involve radio frequencies and most wireless data schemes use packet techniques for transferring data. However, it is the packet radio networks that have become the most closely associated with digital cellular. They are sometimes called packet services because they parcel data bits into specific byte-length packets before sending them into the airwaves at radio frequencies. A special transceiver breaks down the data into packets. It then transmits a stream of these packets into the air. They are picked up by radio towers and forwarded to the proper addressee. Each packet is numbered so that the message can be reassembled at the receiving end. If a packet is not received in good condition, the receiving service automatically asks the sending modem to retransmit the missing or corrupted packet while continuing to receive other packets. In this way, accuracy is ensured and error correction does not delay transmission.

Each of the cellular data technologies has its shortcoming, engendering confusion about which will be the market leader. Most likely, services will coexist with their deployment depending upon the application. Electronic mail and other short messages will travel over the speedier, packet-based network, while longer files and faxes will move along the established AMPS, using modems that are adapted to work in the noisy cellular environment.

Code Division Multiple Access (CDMA)

Digital systems replace FDMA with *time division multiple access* (TDMA) and *code division multiple access* (CDMA) (Table 4.2).

TDMA partitions each radio channel into multiple timeslots and each user is assigned a specific frequency/timeslot combination. CDMA employs a sequence of codes that allows a frequency channel to be simultaneously used by multiple calls. Both systems can support more users per base station per MHz of spectrum than the analog systems that they replace.

TABLE 4.2   Wireless Technology Capacity

Parameter	AMPS	TDMA	CDMA
Bandwidth	12.5 MHz	12.5 MHz	12.5 MHz
RF Channel	0.03 MHz	0.03 MHz	1.25 MHz
Number of RF Channels	416	416	10
Number of RF Channels per cell	59	59	10
Number of Voice Channels per cell	57	171	380
Voice calls per sector	19	57	380
Capacity vs. AMPS	—	3x	20x

CDMA provides three features that improve system quality:

1. It ensures that a call is connected before hand-off is completed, thereby reducing the probability of a dropped call.
2. Variable-rate voice coding allows speech bits to be transmitted at only the rates necessary. This conserves the battery power of the subscriber unit.
3. Multipath signal-processing techniques combine power for increased signal integrity. Additional benefits to the subscriber include increased talk times for hand portable units, more secure transmissions, and special service options such as data, integrated voice and data, and fax services.

Cellular Radio Packet Data (CDPD)

In contrast, CDPD is a digital overlay to an AMPs Cellular Network that allows data as well as voice to be transported. CDPD rides on top of the existing analog cellular infrastructure. It is essentially similar to the packet radio networks in that it transports data in small packets that can be checked for errors and retransmitted as necessary to ensure accurate deliveries. However, because CDPD does this within the current cellular voice network, it uses a technique known as channel hopping to locate idle voice channels and weave data packets into them. CDPD takes advantage of the momentary silences before, during, and after cellular phone conversations to transmit data packets. Hence, CDPD can achieve transfer speeds up to 19.2 Kb/s.

Based primarily on OSI standards, the *common management information protocol* (CMIP) specification defines an architecture that is compatible with the protocols in many existing data networks. At the application layers, it implements X. 400 message handling and CMIP management; at the transport layer it implements a derivative of the IS-IS routing protocol optimized especially for "intermediate" systems. The two basic components of a CDPD connection are the Airlink, which is essentially a radio-based signaling connection, and the network, which is a wire-based supporting infrastructure.

- The Airlink portion is governed by a connection-oriented protocol that provides sequencing and flow and error control. To run error-free, in consideration of the noise of radio channels, most CDPD systems implement new error-correction techniques.
- The network portion of CDPD is basically concerned with connections at the lowest three layers of the network: the Physical layer, the Datalink layer, and the Network layer.

Like a true open system, CDPD supports the Internet IP protocol and the OSI *connectionless network protocol* (CLNP), the two leading connectionless protocols, at the Network layer. Anything above the network layer—APIs, applications, user interfaces, and network operating systems—runs on a CDPD network exactly as it would on a wired network. A unit called the Mobile Database Station provides the relay between the cellular radio system and the digital data. It can transmit data using Frame Relay, X.25, or the Internet Point-to-Point Protocol. Transmissions are sent in short bursts over the same networks used by cellular phones. Mobile Data Intermediate systems recognize the Datalink layer protocols and translate them into cellular data for the end user. At the network termination is a network server or host.

Personal Communication Services (PCS)

Personal communication services (PCS) combine numerous technologies—wireless and wired voice, fax and data telecommunications, computer telephony, and network intelligence—to create untethered communications (Figure 4.1). With PCS there is one telephone number that people can call, regardless of medium (voice, fax, data, etc.) or mode (immediate, deferred, signaling, etc.). Computer intelligence that is hidden behind that number provides a front end for callers, either human or machine, and handles the details of coordinating the media at the back end. Specifically, PCS refers to a portion of the radio frequency spec-

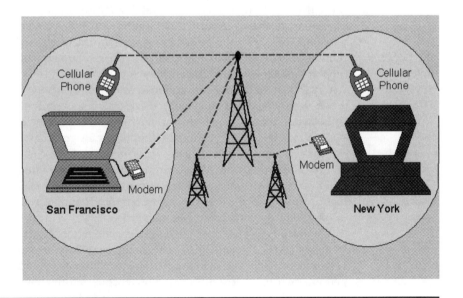

FIGURE 4.1   Mixing data and voice on cellular.

trum (1,900 MHz) which is set aside for shared frequency, small cell, spread-spectrum, CDMA, and CDPD services.

Market research firm BIS Strategic Decisions of Norwell, Massachusetts reports that PCS could fundamentally change the U.S. telephony market. BIS estimates that 151 million adults, or 80 percent of the U.S. total, are interested in reaping the expected benefits of wireless communication. Currently, 95 million U.S. households have conventional telephone service. BIS foresees a future in which a few powerful, full-service carriers compete for wireless business across the nation. Adults who were surveyed expressed an interest in both home and mobile uses for PCS. They also indicated that users will expect such features as single monthly billing and calling party pays. There is a lesser degree of interest in voice mail, message notification, and call screening. Three out of five of those surveyed said that they expected PCS to upgrade or replace their existing telephone or cellular service.

PCS itself is only part of a much larger solution—wireless ramps to the information superhighway for multimedia. In the coming years, PCS based on *advanced intelligent network* (AIN) technology will redefine the way public networks operate.

Wireless LANs

The need for telecomuting has focused attention on wireless LAN technology. More than seven international standards bodies are developing wireless communications standards. The IEEE 802.11 group is defining a common interface for wireless LANs, with the possibility of several distinct physical layer definitions running at data rates from 1 Mb/s to 50 Mb/s. Some see future wireless LANs supporting both voice and data; that is, an IEEE 802 standard digital voice isochronous format for control automation as well as for interactive voice applications.

A variety of wireless technologies are enabling remote users to connect to LANs without telephone lines and allowing LANs themselves to operate without cabling (Figure 4.2). This could represent a significant cost saving for the organizations that use this since a major portion of the cost of LANs is in the cost of cable installation. Moreover, terminals are moved on the average of 1.5 to 3 times per year, further adding to the installation overhead. As older buildings are rewired to comply with strict safety standards, all of this must be factored into the installation planning process.

Wireless LANs use the same bus, ring, and star topologies that are employed by legacy LANs to connect PC nodes in fundamentally the same way that wired networks do; only the media are different. All communications systems must perform two basic actions: They must encode information in a form that is understandable to both the sender and receiver and they must carry the encoded information from source to destination. Digital communications systems encrypt data according to an agreed-upon protocol that usually includes some method of

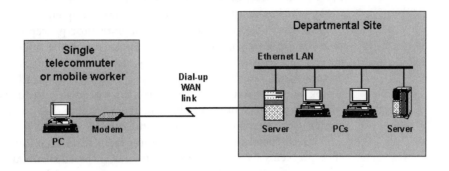

FIGURE 4.2   Mobile user access to departmental site.

error correction. Moving the encoded information without wires relies on transfer techniques that use infrared light, radio waves, or microwave radiation.

Infrared

Infrared signals are light waves at a frequency beyond the range of human vision. Most television remote controls operate using infrared signals. The advantage of infrared is that there is no restriction on its use in the United States or other countries. Infrared systems use a part of the electromagnetic spectrum located just below visible light. Light waves are pulsed from the transmitter to the receiver in either a direct (line of sight) or reflected pattern. With line of sight or focused infrared, a clear channel must be maintained. Infrared light travels in a straight line, so it is blocked by opaque objects and depends upon reflections or repeaters for reaching stations that are not within the line of sight. With infrared LANs, token ring speeds of 4 Mb/s or 16 Mb/s can be achieved with a performance equal to that of wire. If the line of sight signal is temporarily disrupted by someone or something passing through, information is rerouted through the redundant portion of the ring or it is retransmitted when the connection is again established. One way to get around the line of sight limitations is to use flooding, that is the reflected method of infrared transmission. The infrared signal completely floods the room, reflecting light off walls, ceilings, or other objects. Transceivers (mobile or anchored) pick up the signal from the reflected surfaces, then transmit back. While focused infrared can be just as fast as wire, flooded light is much slower.

Spread Spectrum

Radio waves are most often used to interconnect LANs. Whereas network signals are carried on electromagnetic waves for both wired and radio wave

LANs, radio networks usually use frequencies between 902 MHz and 928 MHz, which is about nine times the maximum frequency that coaxial cable can carry. Radio networks require more elaborate error-prevention techniques than do other networks because signals in the air cannot be shielded from interference. To internetwork wired LANs by means of radio waves, a special network driver and an interface card reside in a gateway. The card uses a digital-to-analog converter to translate the LAN's binary signals into analog waves. These analog waves are sent to a radio transceiver, where they then modulate radio carrier waves that are transmitted to the other LAN.

Spread-spectrum radio, a common radio technique, overcomes interference problems by sending broadcasts over a spread of frequencies instead of using a single focused frequency. Spread-spectrum technology operates at the upper end of the UHF band used by television, cellular, and FM radio. Applications using this technology are limited by available bandwidth, although they have the potential for 10 Mb/s transmission. Using spread spectrum, the signal is sent across a broad bandwidth of frequencies to minimize the danger of data loss and the consequent need to retransmit the data, which would slow the network. In one type of spread-spectrum transmission—signal hopping—sophisticated transceivers send data for a few milliseconds on one frequency, then change to another. In a further attempt to reduce data loss, this type of spread-spectrum transmission uses redundancy and sends the same data on several different frequencies in case interference corrupts the signal.

The receiving gateway on the second LAN employs a second transceiver to remove the signal from the carrier wave. An analog-to-digital converter in the network card converts the signal to binary code. In the LAN gateway, the data is pieced together and undergoes statistical checks called *checksums* to ensure that it's uncorrupted. If the checksum at the receiver does not agree with the one generated when the signal was sent, the receiving gateway requests that the data be sent again.

Microwave

Microwave refers to a much higher frequency of radio wave. When a high-speed connection is needed and it is not feasible to connect by wire, microwave is an option. These types of wireless connections are used in situations where networked sites are located within a building or in the same general area but cannot be connected by wire. Examples of the latter are buildings that are separated by roads or railroad tracks. Microwave signals are of a high enough frequency not to be affected by other equipment. They can pass through materials such as office separators and, therefore, are not subject to line-of-sight limitations. Motorola, for one, uses an 18 to 19 GHz band signal for its wireless system. The system consists of a six-sector microwave antenna, a radio frequency digital processor and circuit modules, and high-speed, data-handling silicon integrated circuits.

Wireless Data WAN

Companies are using remote access technologies to connect off-site users into corporate networks. Driven by the prevalence of PCs and the need to link them in a LAN, companies initially built "islands" of LANs at the departmental level and later used bridges and routers to interconnect them to form complex internets. With these interconnected data highways now firmly in place at the enterprise level, the next step involves bringing geographically dispersed users at remote sites into the corporate network via dial-up and remote access technologies. Growth is being driven by the need to access corporate networks from branch offices and by the connectivity needs of mobile workers equipped with laptops, notebooks, and *personal digital assistants* (PDAs). Remote access LAN-to-LAN and client-to-LAN infrastructures have brought nomadic computing into existence, as well as a new class of telecommuters.

There are three primary dial-up approaches available for providing remote connectivity:

1. Terminal emulation. This technique uses software to link a user at a remote terminal across a wide area network to another computer as if it were a locally attached node. Supporting a broad variety of protocols, including TCP/IP, TN3270, DEC LAT, and X.25 PAD, terminal emulation works well for host-based asynchronous applications. Its major drawback is that terminal emulators cannot support client/server applications. Furthermore, the remote device cannot participate as a full peer in the network since it is limited to the functions of a terminal device.

2. Surrogate device. This refers to a device that is directly attached to a local LAN. First introduced in the mid–1980s, remote control software essentially enables a user to take control of a dedicated PC residing on the corporate network. For instance, client/server applications can be supported by running the target application on the surrogate device and the remote control application on both the remote and surrogate devices. The principal deficiency of this approach is that remote control requires each dial-up device to have its own dedicated PC in order to run the application. In addition, remote control applications are implemented using proprietary technologies between a remote control client and server. Finally, remote user access is limited to the LAN applications that the local device can access.

3. Access server. This server supports both client-to-LAN and LAN-to-LAN connectivity over standard telephone lines and public-switched data networks. One of the major implementation differences between remote control and remote access is that remote control systems deal with screen data while remote access systems deal with application data. The remote access server is a hybrid of router and modem technologies that transmits packets of data between systems as if they were directly connected to the LAN. It also has the ability to optimize line use to compensate for the lower band-

width capacity and slower speeds of dial-up lines. Traffic can be aggregated and economies of scale achieved because, like modem pools, incoming lines can be shared across a large number of users.

Wireless Data Services

Wireless data services rely on different types of digital transmission technologies that break down data into ones and zeros that are sent by means of radio waves. Subscribers may receive anything from E-mail messages to lengthy documents on their portable facsimiles, laptop or notebook computers, mobile telephones with a digital read-out window, or *personal digital assistants* (PDAs). Pioneering digital wireless services for two-way transmission include ARDIS, available in about 400 regions around the country and RAM Mobile Data USA, available in 200 large cities. Both use a nationwide system of radio towers to send only text (neither can handle voice) to dedicated hand-held computers. Faxes and other forms of data could be sent over existing analog cellular telephone systems, but they are subject to interference that could cause transmission breaks and other problems.

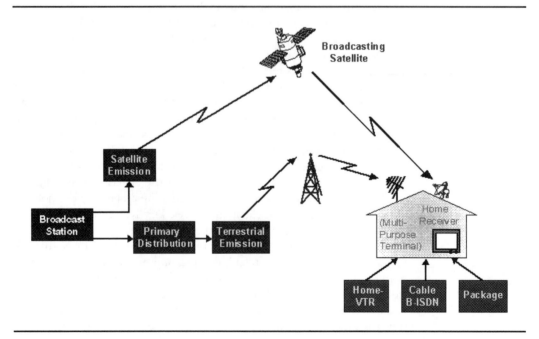

FIGURE 4.3   Broadcast chain distribution.

For very remote wireless networking there are satellite solutions. A satellite connection consists of a dish and transceiver joined to clients or networks at each end (Figure 4.3). Worldwide wireless satellites may prove to be the ultimate information delivery vehicles. And it appears there will be no shortage of celestial trailblazers. Motorola plans to ring the planet with satellites for a global wireless telephone network it calls Iridium, which might be adaptable for data use. Teledesic also aims to put hundreds of low-orbiting satellites around the Earth by 2001. These will be designed to move all types of voice and data. A subscription to one of the satellite services will let people communicate to and from almost anywhere in the world.

CONCLUSION

Eventually all businesses will be connected electronically with a wide variety of information that will be increasingly available for the PC. Information is already being delivered by faster networks with more connections between them, and more services. The creation of ever faster networks is being encouraged by advancements in silicon integrated circuits, fiber optics, and computer software technology. The providers of the services are both traditional and new. At the same time that wireless communications providers continue to be a threat to telephone company local-loop service revenues, the Internet competes with the long-distance carrier business. Although in its infancy, voice telephone conversations made through the Internet via compression software will inevitably take business away from the others.

The data communications equipment industry is experiencing the proliferation of exciting new applications such as facsimile, E-mail, imaging, groupware, and multimedia. As we go forward, societal as well as economic pressures will continue to drive the computer and telecommunications industries in the same direction. Trends based on the expanding of the home work force including telecommuting, coupled with the legislative initiatives that are deregulating and promoting the use of the telecommunications networks, are already creating unrivaled opportunities for broadband technologies and services. These influences will create a global demand for broadband technologies that will change the business productivity equation. Broadband networks will radically alter the boundaries between LANs and WANs as they provide a new format for information transport, switching, and management.

LANs TO WANs

LANs

INTRODUCTION

LANs (local area networks) have emerged as the dominant computing environment, replacing earlier mainframe and minicomputer networks. Initially an assembly of wires and smaller computers, LANs have the ability to connect disparate equipment and that has elevated them to more than mere alternative cabling schemes. The successful IEEE LAN standardization turned Ethernet and token ring into the most common LAN types. Both combine a variety of personal computers, workstations, servers, and peripheral devices into a seamless network. The LAN may attach ordinary telephone station wire to each device on the network and connect it in a wire closet to a backbone coaxial cable between floors as well as to a fiber link between buildings (Figure 5.1).

A new and powerful network element is the server. The server provides the point of interaction between a user and services such as printing and database storage. A variety of computer devices ranging from fast personal computers with large amounts of memory to data switches may be used as LAN servers, provided they are equipped with the appropriate interfaces. The sharing of information and equipment makes LANs relatively inexpensive, allowing them to move from the backwaters to the forefront of networking.

With time, the emphasis has shifted from a single customer premise to the interoperability of multivendor systems encompassing LANs and WANs. Unlike WANs, LANs span a limited distance, often within a single building or group of buildings. But it is becoming less likely for users to be concerned only with their immediate environment. Today, they want to communicate and share resources and information independent of geographic constraints. This encourages the development of better interfaces between LANs and WANs as well as higher data rates.

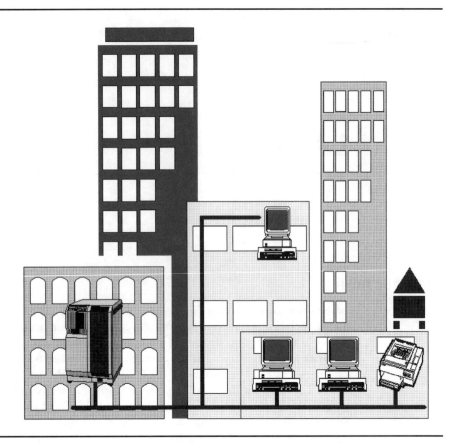

FIGURE 5.1   LANs provide a powerful and efficient solution for campus and local area networking.

LAN OPERATION

The LAN is actually a form of broadband network. It resembles a broadband WAN in terms of raw speed and distance. LANs transport information at rates up to 100 Mb/s and they employ data error protection and congestion control mechanisms similar to those proposed for new broadband WANs. Both frame and cell relay are new WAN technologies that have been borrowed from the LAN. Unlike the conventional WAN, the LAN does not send information in a store–and–forward fashion with error detection and retransmissions. The LAN transports data by means of simple addressing and frame manipulation. The information to be transported is encapsulated within a frame that contains the addresses needed for network routing and management. The LAN protocols arbitrate the shared medium to permit devices on the network to communicate.

To the network architect, the LAN represents an island of interconnected computers. There are two general categories of LAN-related equipment: intra-LAN and inter-LAN. The intra-LAN gear consists primarily of servers and network interface units that provide services and link terminals to the LAN bus. The inter-LAN products—bridges, routers, and gateways— are used to connect different LANs.

LANs differ in terms of topology, access method, connection hardware, media, operating systems, and protocols. Of all of these, topology has the greatest impact upon LAN installation, expansion, and operating costs.

Topology

Network topology refers to the physical and logical layout. The common LAN topologies are bus, ring, and star (Figure 5.2). Each topology influences the performance of a LAN as well as its use in enterprise networks. Unlike a telephone or computer network, LANs have a flat structure that is optimized for device-to-device communication. This is in contrast to hierarchical networks where, for one device to talk to another, the message must traverse the host computer. LANs provide peer-to-peer or client/server computing by means of direct bus, ring, or star connections (Table 5.1), each having its own strengths and weaknesses with regard to speed, reliability, and cost.

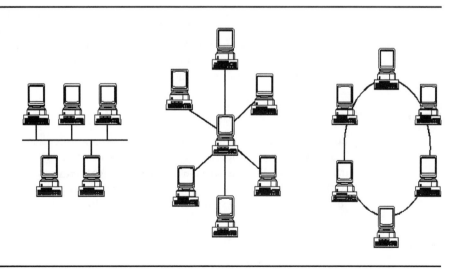

FIGURE 5.2   Bus, star, and ring LAN topologies.

TABLE 5.1   Common LAN topologies

Topology		IEEE standard
bus	802.3	Ethernet
star	802.3	StarLAN
ring	802.5	token ring, ARCnet, FDDI
dual ring	802.6 (DQDB)	MAN

Bus and ring cost efficiencies have made them the dominant LAN topologies. The ability of the ring to provide route redundancy has made it the favorite for broadband networks across which large amounts of information are transported.

Access

LANs employ contention-based and deterministic access methods.

- For contention-based LANs such as Ethernet, all stations contend for available bandwidth on a first-come, first-served basis. If two or more stations access the LAN at the same time, a "collision" occurs. Whenever a collision is detected, all stations back off and try to gain access to the bus at staggered intervals (Figure 5.3).
- Token ring employs a deterministic method of access. Each station is allocated a time interval during which it is guaranteed access to the shared ring. The time to transmit is controlled by a continuously circulating token. When a terminal needs to transmit, it replaces the token with packets of data. After the transmission is completed, the token is reinserted, allowing another terminal to transmit.

Connection Hardware

Ethernet LAN devices attach to the bus by means of a *network interface card* (NIC) that is inserted into a computer or peripheral device. The NIC cable attaches to a transceiver, which plugs into the bus cable. For token ring, the connection is made with an adapter card that is inserted into each computer and peripheral device. The adapter attaches to the *media access unit* (MAU) which plugs into the ring cable.

Media

The removal of media constraints has accelerated wireless LANs' popularity. LANs operate over a variety of materials, ranging from voice-grade, twisted-pair wiring to

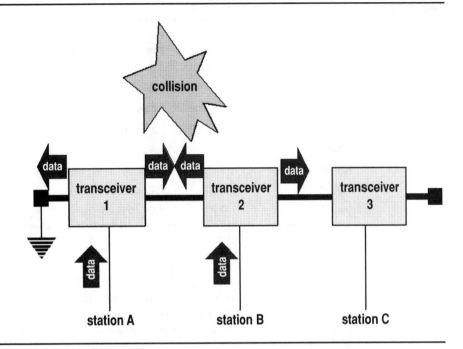

FIGURE 5.3   Ethernet collision.

fiber cable to radio waves. The anticipated data rate determines the media. Early LANs used thick coaxial cabling that was inflexible and difficult to install. Later, more flexible thin coaxial cabling was introduced and lower-cost twisted-pair wiring followed. Twisted-pair wiring already exists in most office environments because it is used in telephone systems. Backbone LANs with data rates of 100 Mb/s use fiber optic cabling. Fiber optic cable also improves immunity against *electrome-chanical interference* (EMI) and *radio frequency interference* (RFI). Wireless LANs use radio waves to connect terminals on the same floor of a building. Although wireless LANs are not as secure as those that are wired, they can be improved by employing spread-spectrum transmission and encryption algorithms.

Protocols

Any base-layer LAN protocol can be replaced by another without the user being aware of it. What does change is the LAN performance, particularly when one goes from the prevailing twisted-pair to fiber optic media (Table 5.2). With LANs, performance is measured by both speed and cost. LANs employ simple protocols that, while fast and inexpensive to implement, only

TABLE 5.2   LAN protocol performance.

Technology	Stated bandwidth (Mb/s)	Real bandwidth (Mb/s)	PC card cost (relative)	Media cost (relative)
Ethernet	10	1.5 to 3.0	low	medium
token ring	16	11 to 13	medium	medium
FDDI	100	70 to 80	high	high
FOIRL	10	1.5 to 3.0	medium	high
Fast ethernet	100	15 to 30	medium	medium
ARCNET	2.5	1.7 to 2	low	medium
LocalTalk	0.25	0.04 to 0.08	no charge	low

work over limited distances. Unlike WAN protocols such as X.25, they operate at the data link or frame level and therefore lack the extensive error protection required for long-distance transmission over noisy copper lines. Now, the vastly increased amount of relatively noiseless fiber cable installed over the past decade in the public telephone network and the increased intelligence in the end-user terminals, are allowing these simpler protocols to migrate into the WAN.

Network Management

The interface that the user sees and the number of users that share the network at the same time are governed by a network management system. LAN network management is responsible for more than just provisioning and maintenance. These systems incorporate security and data protection, as well as databases for resource accounting and services such as E-mail. The management system uses software agents that reside in each device on the LAN. Instructions are communicated to individual devices from a central network management station.

When LANs were separate islands of computing power, there was no need for the scope of the management station to extend beyond the LAN on which it resided. Independent LANs are being replaced by LAN internets which require that their managers control the remote devices that are on other LANs. Each LAN can have its own local manager and yet be subservient to a remotely located management station. The situation is made even more complicated by the existence of other types of management systems, such as mainframes and WANs. The need to work in a hierarchical management environment has created the need for standardized ways to handle fault, security, configuration, performance, and device management. Various management tools include:

- Fault management tools identify a network problem with little or no human intervention. They forward alarms to other devices in the network or to the network administrator. These sophisticated tools have the ability to learn from prior faults. When a fault occurs, a database is searched so that possible solutions can be forwarded with the alarm.
- Configuration management tools deal with the physical network properties. These tools identify each network element such as bridges, routers, terminals, and so forth. This information is stored in the network database that is used to construct a map of the network topology with all of its elements.
- Security management tools ensure system integrity. Some tools can configure a single address that receives or transmits data at the port level. When another address tries to communicate through this port, it will automatically be partitioned or blocked from entering the network. This allows the creation of secure subnetworks that protect the integrity of confidential information.
- Performance management tools allow network statistics to be viewed and collected. Alarms are generated when performance falls outside established limits.
- Element management protocols, such as *simple network management protocol* (SNMP), allow disparate devices and management systems to communicate. SNMP is the most widely used way to gather and manage information and, if required, can read an index of register values within a network device and change them. Associated with SNMP are a number of *management information bases* (MIBs) that assign a standard meaning to the values. Over time, several values have developed standard meanings, but special functionality is still handled by proprietary MIB extensions.

Distributed management is particularly important for broadband LANs and internets where there may be thousands of users. The resulting information is valuable to the staffs at all levels in an enterprise. Should they want to monitor alarms and trouble tickets, a distributed management system allows this information to be forwarded to all clients and servers that request it, thereby keeping everyone up to date. A critical alarm, for example, could be passed from the alarm collection server to the trouble ticket database server. There, a problem record could be created and propagated throughout the network.

BROADBAND LANs

Although the management of heterogeneous LANs requires sophisticated systems, the broadband LAN itself is a relatively simple network. LAN architecture may be separated into two parts: the *media access control* (MAC) and the *logical link control* (LLC), sublayers of the OSI data link and physical layers (Figure 5.4). The MAC sublayer specifies how a device transmits and controls the signal over transmission media that ranges from coaxial cable, twisted-pair

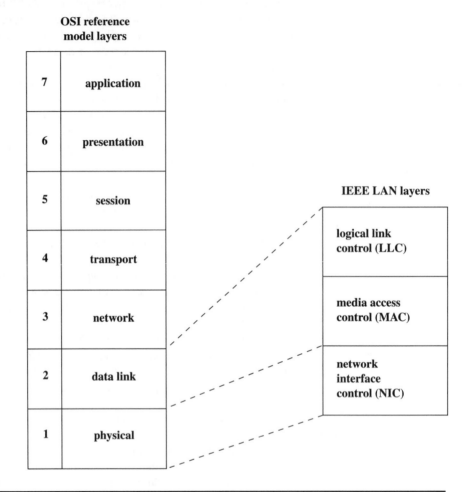

FIGURE 5.4   IEEE standards relation to OSI model.

wiring, and fiber-to-radio frequency. A variety of medium-dependent access control methods are standardized by the IEEE, including *carrier sense multiple access with collision detection* (CSMA/CD), token ring, and FDDI. The upper sublayer, LLC, adds the routing capability to the data link layer. In this respect, the IEEE standard differs from the ITU-approved OSI model. The OSI model relegates routing to layers 3 and 4. The transport of frames, like packet-switching, requires some form of addressing to distinguish which data belongs to individual users. Connectionless and connection-oriented routing are possible with the LLC, which establishes the connection, transfers data, and then terminates the connection.

There are three ways that LANs route information:

1. Unacknowledged connectionless service: For transport within the LAN there is no logical connection between source and destination. The frames are delivered on a best-effort basis, employing datagram service. Delivery is not guaranteed and lost frames are simply dropped, requiring the receiving device to request that the frames be retransmitted.
2. Connection mode service: For interconnecting LANs, a logical connection between source and destination is established before transmission. This improves the efficiency of lengthy exchanges while relieving higher-level protocols from the burden of connection management.
3. Acknowledged connectionless service: This is used for specialized applications such as point-of-sale or factory assembly where a large number of limited intelligence devices may communicate with a central processor. Frame receipt is acknowledged by the data link layer.

By avoiding the processing overhead incurred by higher layer functions, LAN protocols are fast and relatively easy to implement.

The architecture of the LAN has always been vastly different from that of the WAN. To begin with, the LAN uses protocols that would not work over the long distances associated with the WAN. These protocols give the LAN its broadband attributes, while maintaining a low-cost medium to interconnecting computers. In contrast, WAN communications protocols ensure that information is protected from errors during transmission. The media employed by LANs also differs: for example, unshielded twisted-pair, coaxial cable, and multimode fiber cable are not used in WANs. Finally, the nature of the transported traffic differs—LANs were developed to transport data, WANs to transport voice. As a result, the standards, equipment, and functions of the two types of networks are so different that people think of them as separate entities.

With the increasing importance of broadband communications, these differences are being revisited, primarily because the LAN is the most successful form of broadband network. Inherent LAN features such as frame-level communications were adopted by WAN technologies such as frame relay. And computer-derived technologies such as ATM, initially developed for the WAN, are entering LANs.

LANs come in many forms and varieties. This chapter concentrates on three of the standard ones: Ethernet, token ring, and FDDI (Figure 5.5).

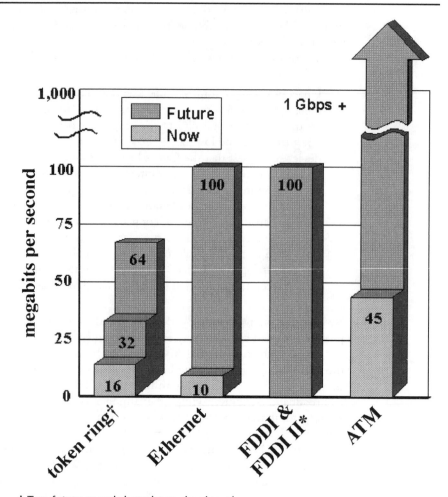

† Two future speeds have been developed
* No increase is expected
Source: Communication Week

FIGURE 5.5   Current and potential future speeds of various LAN types.

Ethernet

With millions of networks in place, Ethernet is the most widely deployed LAN based on a bus topology (Figure 5.6). By itself, it accounts for over 45 percent of the installed LAN nodes. In the early 1970s, Ethernet was promoted by a coalition comprised of Xerox, DEC, and Intel (Xerox was its inventor, DEC

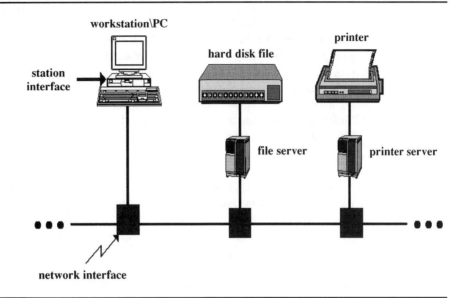

FIGURE 5.6   Ethernet local area network.

the market provider, and Intel the silicon integrated circuit supplier). Ethernet allows all computers on the network equal access to the linear bus. Information transported by Ethernet is reduced to relatively small frames that include source and destination addresses as well as error-protection mechanisms. The architecture was designed to transfer data at 10 Mb/s, maintaining simplicity and low cost. Trade-offs in the way that collisions are detected and the traffic is controlled limit the distance over which the protocols are effective.

The access method, CSMA/CD, regulates how terminals share the common bus. The probability that a collision will occur depends on the number of terminals on the LAN: the more terminals, the greater the number of collisions. Each terminal "listens" to determine if the bus is idle before transmitting its frame. If several terminals attempt to transmit at the same time, a collision results and the data becomes distorted. This condition is resolved by having the terminals release the bus and reconnect at staggered intervals. Another source of delay is the way that traffic volume is controlled. When there is too much traffic for a receiving device, the device simply discards the message. Since messages are numbered and the transmitting device waits for an acknowledgment, discarded messages can be detected and retransmitted, but this substantially degrades performance. In addition, some applications cannot tolerate the lengthy delays that occur with CSMA/CD.

Ethernet uses three types of cables; thin coaxial, thick coaxial, and unshielded twisted-pair—respectively referred to as 10BASE2, 10BASE5, and 10BASE-T.

The most recent standard, 10BASE-T, works with existing telephone wiring systems. The use of CSMA/CD imposes a practical limit on the length of the bus. The IEEE 802.3 Ethernet standard allows 500-meter cable lengths without repeaters for signal regeneration. Repeaters extend the bus to 2,500 meters—approximately 2.5 miles. A maximum of four repeater units may be in the signal path between any two stations on the network. While repeaters amplify and reconstitute weak signals, they do not compensate for signal propagation delays. Unless distance limits are observed, signal delays may become so great that the terminal at the end of the bus may not detect that another terminal is transmitting.

Frame

A data stream represents a pattern of binary bits. With Ethernet, this stream is transported as frames (Figure 5.7). That is, data is encapsulated in an IEEE 802.3 standardized format called a frame that is recognized by any Ethernet adapter. The frame contains addressing, routing, and error-checking information as well as the following data fields:

1. Preamble field: Each frame begins with an 8-byte preamble field: 7 bytes are used for synchronization and to define the frame. The remaining byte indicates where the frame begins.
2. Address Field: The address field consists of destination and source addresses. The destination address field identifies the location of the receiving terminal. The source address field identifies the sending station. The address field can be either 2 bytes (16-bits) or 6 bytes (48-bits) in length. A destination address can refer to one terminal, a specific group of terminals, or all terminals.
3. Control Field: The Ethernet control field indicates the length of the data field and provides padding for detecting collisions. The length of the data field is indicated by the 2-byte length count field. This field determines the

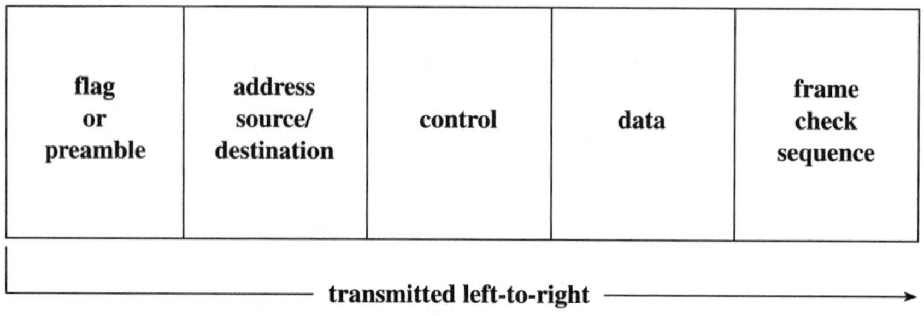

flag or preamble	address source/ destination	control	data	frame check sequence

————————— transmitted left-to-right —————————→

FIGURE 5.7   Generic IEEE LAN frame.

length of the data unit when a pad field is included in the frame. Padding is used for collision detection. A frame must contain a specified number of bytes. If a frame does not meet this minimum length, extra bits are added.

4. Data Field: Transported information is encapsulated in the data field as 8-bit bytes. The minimum frame size is 72 bytes; the maximum is 1,526 bytes. The maximum frame size reflects practical considerations related to adapter card buffer sizes and the need to limit the length of time the medium is tied up in transmitting a single frame. If the data to be sent exceeds 1,526 bytes, the higher layers break it into individual packets in a procedure called *fragmentation*.

5. Frame Check Sequence: The frame check sequence terminates the frame. It has two purposes: to define the end of the frame and to check for errors. Both the sending and receiving terminals perform cyclical redundancy checks on the frame bits. The sending terminal stores the result of this calculation in the 4-byte frame check sequence field. The receiving terminal compares the cyclical redundancy checks that it calculates with this value. If the two numbers do not match, a transmission error has occurred and the frame is retransmitted.

Network Addresses

One reason for the popularity of IEEE-standard LANs is the unique address given to every device that connects to the LAN. Ethernet supports universal and network-specific addresses. Manufacturers of Ethernet NICs are granted address space that they alone can use, ensuring that every Ethernet device has a unique and identifiable address no matter where it is located. (The maximum source or destination address is 2(48).) Thus, network interconnection devices such as bridges can automatically route Ethernet frames based on the existence of a unique address for the receiving terminal. Whether universal or network-specific, the address can be set by the terminal itself during initialization. Any frame sent to this address is received and processed by the terminal. With universal addressing, all devices on the network have unique addresses. With network-specific addressing, each terminal is given an address that is unique within the network, but which can be the same as a terminal on another network. Because Ethernet does not specify how the 48 bits of an address must be used, network-specific addressing is possible. In this case, when networks are interconnected, a unique network identifier is attached to the terminal address to provide a unique address.

Ethernet also supports the use of multicast and broadcast addresses. An address consisting of all "1" bits is defined as the broadcast address and is received by all terminals. An address associated with a particular group of stations is a multicast address. A multicast address is identified by the value "1" in the first bit of the address. Individual terminals can be enabled for multicast, allowing terminals to accept frames with multicast addresses.

Physical Elements

Ethernet defines the following electrical and mechanical characteristics that enable components from different vendors to be interconnected.

- Physical configuration: This defines limits, cable length, number of repeaters, total path length, and transceiver cable length. The most common Ethernet architecture, 10BASE5, uses baseband transmission over coaxial cable at a data rate of 10 Mb/s. The maximum cable length is 500 meters.
- Coaxial cable specifications: This includes the cabling, connectors, and terminators. 10BASE5 installations use a relatively expensive 50-ohm coaxial cable with a diameter of 10 mm, now referred to as *thick Ethernet cable*. Another cable standard, 10BASE2 (10 Mb/s, baseband signaling, 200 meters), uses ordinary CATV-type coaxial cable, called *thin Ethernet cable*. Twisted-pair wiring has emerged as an alternative under the 10BASE-T standard.
- Transceiver specifications: This includes the cable as well as the transceiver. The migration from thick coaxial cable to thin Ethernet (*Cheapernet*) to telephone twisted-pair wiring increases the complexity of the transceiver, but saves both installation and cable costs.
- Environmental specifications: This includes temperature, humidity, etc.

Ethernet has demonstrated a surprising ability to adapt to the market demands for greater economy and speed. Some vendors have suggested increasing Ethernet's 10 Mb/s capacity with an extension to existing standards; others want to boost the performance of existing Ethernet LANs through proprietary methods.

Fast Ethernet

Multimedia applications are pushing Ethernet LANs to their limits. Conventional Ethernet performance is limited in several ways. First, the data rate and contention protocols do not support voice traffic. Second, a shared-media LAN limits the amount of bandwidth available for specific applications. As more users are added, each receives a smaller percentage of the total bandwidth. The rate at which a single device—server, personal computer, or workstation—can transmit and receive from the network is limited because they all share access to the bus. For example, if a LAN has two servers, each with the capability of transmitting at the full Ethernet bandwidth, the throughput of each server is halved.

One temporary solution to the bandwidth crunch is Ethernet switching, which dynamically allocates the full 10 Mb/s connection to each user on the Ethernet network. Another answer is full duplex-switched Ethernet that allows nodes to simultaneously transmit and receive 10 Mb/s per switch port or an effective throughput of 20 Mb/s. However, as LAN traffic continues to rise, these become temporary fixes and network operators must turn to other faster technologies such as ATM, FDDI, and fast Ethernet. Fast Ethernet offers the least radical enhancement.

The term "fast Ethernet" describes four variants—100Base-T4, Tx, Fx, and 100VG-ANYLAN (which supports both fast Ethernet and fast token ring). Despite the number of variations the basis is still Ethernet, rather than a new technology; consequently, it provides an evolutionary rather than revolutionary pathway away from legacy networks. Because it uses existing equipment, it is relatively cost-effective to implement. Moreover, it is technically superior: Conventional Ethernet transmits data at 10 Mb/s, while fast Ethernet accommodates transmissions of 100 Mb/s.

There are many other benefits to fast Ethernet:

- Easy Migration: Because the core of 100Base-T technology (the CSMA/CD media access control layer) is virtually unchanged from 10 Mb/s Ethernet, 100Base-T is easily introduced into standard Ethernet environments.
- Proven Technology: With over 30 million Ethernet nodes in use, 10 Mb/s Ethernet technology has proven to be reliable, robust and low-cost. Users, MIS managers, and systems integrators are already familiar with the technology, cabling, and software used in fast Ethernet networks.
- Multivendor Support: Most networking vendors offer products that work with those of other vendors.
- Low Cost: Fast Ethernet will retain the traditional cost advantages that standard Ethernet has held over non-Ethernet technologies.

The low cost and ease of migration provided by fast Ethernet make a compelling argument for upgrading today's networks to higher-speed technologies such as ATM, when they become too slow for critical applications.

Impact on Legacy LANs

One of fast Ethernet's strengths is that it can be used with shared-media and switched 10Base-T networks, allowing any mix of fast Ethernet and conventional Ethernet networks. The 100Base-T design uses switches to link small, easily managed networks, which provide service to users while linking servers over 100 Mb/s links. Switching can be used to ratchet service down to 10 Mb/s or to provide a collapsed backbone to connect multiple 100 Mb/s links. Fast Ethernet is well suited to a corporate environment where network administrators can configure mixes of legacy 10 Mb/s and 100 Mb/s segments. For example, a network supporting finance, sales, and Research & Development could use a combination of inherited 10 Mb/s and new 100 Mb/s segments (Figure 5.8). The finance department, which has only occasional network demands, uses the legacy 10 Mb/s Ethernet; while the sales department with more volume uses switched Ethernet; and Research & Development (R&D), with the

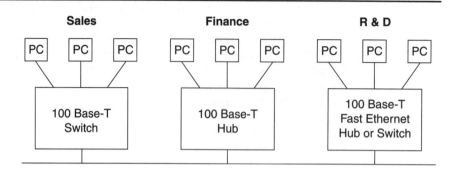

FIGURE 5.8   Combined Ethernet network.

most traffic, uses fast Ethernet, which may be switched for more bandwidth. Fast Ethernet is able to overlay existing Ethernet networks because it supports most in-place wiring.

Implementation

Ethernet popularity is based on its cost-effective connectivity. Fast Ethernet follows this tradition. Like legacy Ethernet, fast Ethernet is a shared–media technology: Client nodes connect to hubs or repeaters and the 100 Mb/s bandwidth is shared among all users. A key ingredient of fast Ethernet is the use of existing copper wiring as well as fiber-optic cabling. The IEEE standard defines four OSI physical layer specifications in order to support different cabling types:

1. 100Base-T4 provides 100 Mb/s Ethernet over four pairs of Category 3, 4, or 5 *unshielded twisted-pair* (UTP) cable. It uses a ternary 8B/6T encoding scheme in which 8 binary bits are encoded in 6 three-valued symbols (-1, 0, +1). These symbols are divided among three pairs of wire for data and the fourth wire pair for collision detection, providing a symbol rate of 25 Mb/s. It can only operate in half duplex: Two pairs are used for unidirectional transmission and two for bi-directional transmission. Moreover, 100Base-T4 cannot be used in the 25 pair wiring found in some older buildings because of susceptibility to *near end crosstalk* (NEXT).

2. 100Base-TX provides 100 Mb/s service over two pairs of Category 5 UTP cable using the TP-PMD physical layer signaling system developed for transmitting FDDI over copper. One pair transmits and the other pair receives signals. With a clock rate of 125 MHz, its 4B/5B encoding yields an effective bandwidth of 100 Mb/s. It can run either full duplex (receive and transmit) over two pair or half duplex (receive or transmit) over four pair. The length of 100Base-T links can be extended to 2,000 meters by

using full duplex; whereas, half duplex copper links are limited to 100 meters and fiber links to 412 meters. A drawback of 100Base TX is that its high clock rate over UTP makes it harder to satisfy FCC Class B emission standards than with other variants. Also in cable plants based on Category 3 or 4, 100Base-TX requires a cable upgrade. Nonetheless, of the four specifications, it is the most popular.

3. 100Base-FX provides 100 Mb/s service over two multimode fibers, again using FDDI signaling systems. It has zero electromagnetic emissions. Otherwise, it is the same as 100Base-TX.

4. 100VG-ANYLAN uses 100 Mb/s transmission over Category 3, 4, or 5 UTP, Type 1 *standard twisted-pair* (STP), or fiber-optic cables. All four pairs of wires transmit and receive data, enabling 100VG-ANYLAN to operate at a signal rate of 25 MHz, resulting in the 100 Mb/s network operating rate. Data scrambling and encoding reduce crosstalk, allowing the use of 25 pair wiring. It supports half duplex transport over Category 3, 4, or 5 wire and full duplex over shielded twisted-pair and fiber cable. The cable rules are somewhat different for 10 Mb/s or 100 Mb/s Ethernet. Node-to-node or node-to-hub cabling distances are 100 meters for Category 3 and Type 1 cable, 150 meters for Category 5 cable, and 2,000 meters for fiber-optic cable. Maximum network span is 4,000 meters; up to four concentrators are allowed between any two nodes. Up to five repeaters may separate any two 100VG-ANYLAN end nodes with up to 100 meters between one device and another. In contrast, other fast Ethernet variants are limited to two repeaters between the end nodes.

Standards

Fast Ethernet standards are driven by industry groups such as the Fast Ethernet Alliance, individual companies, and the IEEE. Although the market has not decided which standard will dominate, there are some hints. In the past, standards that were technically superior gave way to less sophisticated standards because they did not offer an economical migration path. NTSC television is a case in point where an inferior technology won on the basis of backward compatibility with the enormous installed base of black-and-white television sets. If this applies today, then the IEEE fast Ethernet standard 100Base-T should dominate with 100VG-ANYLAN relegated to niche applications.

100Base-T (IEEE 802.3)

By preserving Ethernet's core specification, the CSMA/CD MAC significantly reduced the time to standardization since Ethernet users are already familiar with the technology, cabling, and software used in 100Base-T networks. The standard operates over ordinary voice-grade unshielded twisted-pair copper wiring as well as fiber cabling so that users have the option of rewiring fast Ethernet. For the physical layer, it uses the *American National Standards Institute*

(ANSI) X3T9.5 *physical medium-dependent* (PMD) sublayer and the 100Base-T variants use the existing Ethernet MAC. The core of 100Base-T technology, CSMA/CD media access control layer, is virtually the same as 10 Mb/s Ethernet, which makes 100Base-T easy to introduce to standard Ethernet environments. 100Base-T retains the collision-based CSMA/CD scheme used to arbitrate access to the Ethernet wire and the same 802.3 frame format and CSMA/CD architecture used in conventional Ethernet. Nonetheless, it requires users to upgrade network adapters and to segment the network with routers, bridges, and switches.

There is also a *media-independent interface* (MII) for fast Ethernet that allows it to support various cabling types on the same network. The MII interface is similar to the *attach unit interface* (AUI) found on most legacy Ethernet adapters and supports interchangeable media interfaces through internal or external transceivers. These transceivers attach to the MII connector and handle the various types of Ethernet cables.

While a network is in transition, configuring the mix of 10 and 100 Mb/s full- and half-duplex switches, hubs, and adapters can be simplified because the fast Ethernet standard includes an optional configuration auto-detection mechanism *Nway*. A hardware implementation on the NIC, Nway detects whether a switch, shared-media hub, or another network adapter is capable of 100 Mb/s communication. If not, it defaults to 10 Mb/s legacy Ethernet.

100VG-ANYLAN (IEEE 802.12)

100VG-ANYLAN is a 100 Mb/s protocol developed by a consortium that includes Hewlett-Packard and AT&T. The 100VG-ANYLAN technology has some promising capabilities, but is not as widely supported as the fast Ethernet standard. It supports standard 802.3 Ethernet and 802.5 token-ring frame types, but uses a completely different method of arbitrating access to the wire, replacing CSMA/CD with a deterministic scheme called *demand priority access* (DPA) which is used to arbitrate access to the wire. DPA is similar to token ring in that it increases bandwidth by eliminating collisions and retransmissions. In operation, a station requests permission to transmit data at a specific level of priority. A concentrator or hub scans each port in succession, thereby eliminating the collisions—and the transmission delays—that often occur in heavy-traffic CSMA/CD networks. The concentrator grants permission to transmit and directs the incoming data to the appropriate destination. If multiple transmissions occur at once, the higher-priority requests are serviced first. The demand protocol can supply guaranteed bandwidth and priority of service, two attributes critical for multimedia networking. Guaranteed bandwidth is also important for scaleability because, under the current 10 Mb/s Ethernet standard, the performance of the LAN declines as more users are added. With the demand protocol, the performance of the network is bounded by the speed of the concentrator.

100VG-ANYLAN, an alternative technology to 100Base-T, is designed for applications that 100Base-T does not support. Specifically, 100VG product features (e.g., demand priority queuing) allow 100VG to prioritize support for low-latency applications such as interactive video or distributed engineering programs. The FDDI/token ring-like architecture of 100VG, which polls all stations in the network for traffic, also enables 100VG to maintain a very graceful degradation under high traffic loads. In this light, 100VG is probably more of an FDDI replacement than a 100Base-T competitor.

Token Ring

Token ring is a token-passing network that operates at 4 or 16 Mb/s (Figure 5.9). Information exchange occurs via a token that is circulated around the ring. Each terminal, in sequence, is given a chance to put information on the network. When a terminal accepts the token, the token is replaced with a frame containing data. The destination terminal retains the message and only the

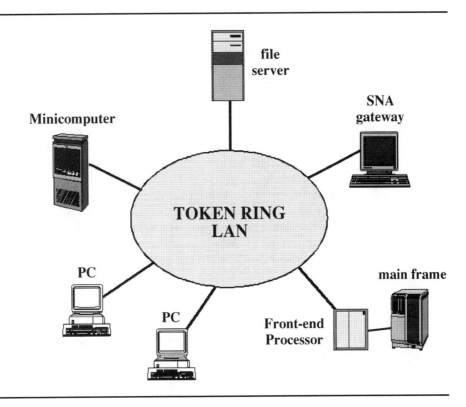

FIGURE 5.9   Token ring LAN.

station that put the message on the ring can remove it. The maximum amount of time that a terminal occupies the network before passing the token to another terminal is determined by a token-holding timer. Because each terminal regenerates the token back to the original signal strength, ring LANs support greater distance and speed than bus LANs. Ring networks also provide flow control, which is implemented with a *refuse bit* near the end of the message. The receiving terminal can leave this bit set, indicating to the transmitter that the message was not accepted.

The ring topology has been reserved for LANs in which users are not concerned about the cost of duplicate lines. Currently, this topology is migrating to broadband WANs as WAN users come to appreciate the ring's advantages, which are:

- Higher data rates
- Fewer routing problems since messages may be broadcast to selected nodes
- Simplified control because little in the way of additional hardware or software is needed
- Improved reliability because there is path redundancy
- Ability to grant preference to high priority traffic due to a priority indicator that is embedded in the token

Specific procedures such as *neighbor notification* ensure that new terminals are recognized by others and granted a proportionate share of network time. Upon power-up, each station becomes acquainted with the address of its neighbors on the network. Addresses are periodically rebroadcast thereafter. A special frame is broadcast to all of the stations on a ring. The first terminal that is downstream of the broadcaster will recognize that certain bits within the frame are zero. This terminal has now learned the address of its nearest neighbor and resets some of these bits to "ones" so that terminals farther downstream will not see all the bits as zero. This process continues in a daisy chain fashion until every station knows the identity of its upstream neighbor. This knowledge is important in case of a ring failure. When a hard failure occurs because of a broken cable, jabbering station, or failed PC adapter card, among other reasons, its cause is isolated by the upstream station which reports a failure by transmitting beacon MAC frames.

Token ring LANs are not without problems. Failed nodes and links can break the ring, preventing all of the other terminals from using the network unless a dual ring configuration with redundant hardware and bypass circuitry is used to isolate faulty nodes from the rest of the network. With bypass circuitry, physically adding or deleting terminals from the token ring network can be accomplished without breaking the ring. Token rings are also vulnerable to terminal failures that occur before the terminal passes the token. The network is then down until a new token is inserted. The token may also become corrupted to the point of being unrecognizable to the terminals. When this occurs, a token timeout alerts all stations on the ring that the token protocol has been suspended. The network can also be disrupted by the occasional appearance of two tokens or by the presence of a continuously circulating data packet. This happens when data is sent to a failed terminal and the originating terminal becomes disabled before it can remove the packet from the ring. These failures require more sophisticated recovery mechanisms.

To protect the token ring from potential disaster, one terminal is typically designated as the control station. This terminal supervises network operations and does important housecleaning chores such as reinserting lost tokens, taking extra tokens off the network, and disposing of "lost" packets. To guard against the failure of the control station, every station is equipped with control circuitry so that the first station that detects the failure of the control station assumes responsibility for network supervision. A variation of token-passing allows devices to send data only during specified time intervals. The ability to determine the time interval between messages is a major advantage over CSMA access methods. Since each device transmits during only a small percentage of the total available time, no one station uses the full capacity of the network. This time-slot approach can support voice transmission and may be employed more frequently as token ring LANs adopt 100 Mb/s data rates.

THE FIBER DISTRIBUTED DATA INTERFACE (FDDI)

FDDI is a high-performance, 100 Mb/s backbone LAN supporting high-powered workstations and slower local area data networks (Figure 5.10). It was the first truly broadband LAN. There are thousands of FDDI LANs installed and ANSI and OSI standards for these are well developed. The ANSI standard for FDDI defines two parallel, timed token-passing rings: one active and the other in backup for a secure physical link. An FDDI LAN can support up to 1,000 physical connections over distances of 200 kilometers. Originally, the FDDI standard specified the use of multimode fiber, which limited the maximum distance between two nodes to 2 kilometers. Use of single-mode fiber has extended this distance to up to 40 kilometers (Table 5.3). By this means, FDDI can stretch over greater distances and provide higher bandwidths than Ethernet or the token ring LANs. It is a solution for such leading-edge applications as:

- Archiving: FDDI can minimize the time required to back up files.
- Network backbone: Low-speed LANs may be joined using an FDDI LAN. For Ethernet, this provides 10 times the bandwidth.
- Computer room networks: An FDDI LAN may locally connect high-speed computers.

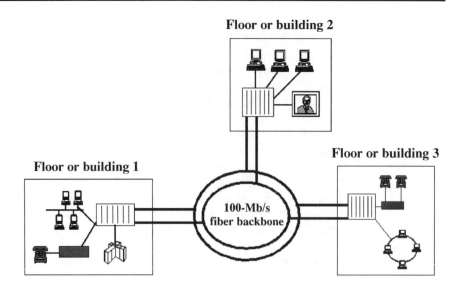

FIGURE 5.10   FDDI merges data, video and voice.

- High-speed LAN: An FDDI wiring hub may be connected in a star configuration to FDDI-equipped computer workstations.
- CAD/CAM: Large CAD/CAM images can be transported, allowing instantaneous sharing of files between multiple stations.
- File servers: File servers can be connected directly to an FDDI backbone, thereby speeding the flow of information.
- Image processing: Real-time access to large graphics files is vital for such applications as medical imaging.

TABLE 5.3   FDDI Media vs. Distance.

Media	Distance
Multimode fiber	up to 2 Km
Singlemode fiber	up to 40 Km
Twisted-pair copper (TWP–150 ohm)	up to 100 meters
Unshielded twisted-pair (UTP–100 ohm)	less than 100 meters
Thin-wire coaxial cable (RG58–50 ohm)	up to 100 meters

Operation

FDDI uses a token-passing access method—similar to token ring LANs—over two counter-rotating optical fiber rings. One ring is set up as the primary ring, the other as the backup. The dual-ring topology provides redundancy; if there is a failure on the primary ring, the backup ring automatically begins transporting traffic. On an FDDI ring, stations attached to both rings are called *dual access stations* (DASs) and stations attached to just one ring are called *single access stations* (SASs). Unlike DAS, links between SAS nodes are not self-healing since there is only one ring between them. FDDI allocates bandwidth both asynchronously and synchronously. Synchronous transmission allows continuous, fixed, data-rate conversations. Asynchronous dialogs allow two terminals to keep the token and exchange data between themselves for extended periods.

The four key components of FDDI are:

1. PMD: At the lowest level, the PMD handles the electronic transition. It converts optical signals from the fiber into electronic pulses and passes the bit information to the physical layer.
2. Physical protocol at the physical layer: Timing and encoding take place in the physical layer.
3. MAC at the data link layer: The MAC layer assembles bits into frames that are handed to the network via IEEE 802.2 logical link control (LLC).
4. SMT: Transcending both layers, the SMT allows FDDI stations to communicate with each other for connection, configuration, and fault management.

SMT also defines the management information base (MIB) in each station, which includes a set of managed objects and their attributes. A MIB might contain the number of ports in the station, its current configuration, and statistics such as the number of frame errors. The SMT software also defines operations that can be performed on managed objects and the way in which events and conditions are handled. Although FDDI is significantly more expensive than Ethernet or token ring, the costs may be justified by its increased speed, reduced downtime, and fewer backbone congestion problems. Moreover, its network management features are built in. Therefore, changes in configuration are implemented transparently to the user, providing a fault-tolerant and self-healing system (Figure 5.11).

There are variations of FDDI that lower installation and operating costs. For example, *copper distributed data interface* (CDDI) uses copper wire rather than fiber-optic cable. However, while reducing costs, CDDI is not widely deployed because of limitations including: shorter distances between stations (100 meters or less); susceptibility to electromagnetic interference (from sources such as electrical motors); and fiber-optic DAS connections needed for network redundancy.

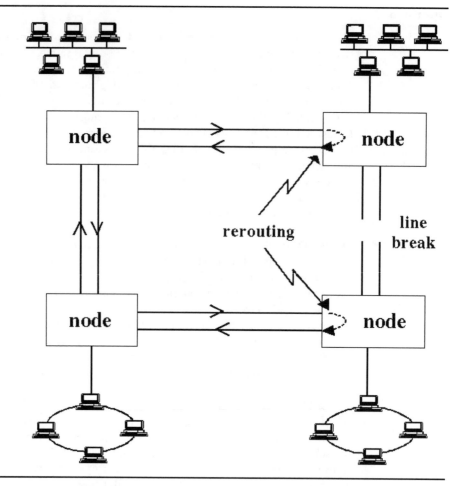

FIGURE 5.11   FDDI self-healing.

Voice on FDDI-II

FDDI extensions transport voice as well as data. FDDI transmits data frames but does not guarantee when the frames will be delivered. Voice transmission, however, requires an isochronous transport capable of guaranteeing access to the network at specified time frames. Using an isochronous service is like making a telephone call: One makes a reservation for the length of the call and the amount of bandwidth that is needed. Connection delays are small enough so that data and voice transport can be transported together.

To support isochronous transmission, FDDI-II extends the FDDI specification with an overlay of synchronous services. Connections between nodes can be made for guaranteed duration and data rates. FDDI-II uses updated versions of MAC and the physical layer known as MAC-2 and PHY-2. SMT-2, an expanded version of the SMT, includes the services needed to support isochronous circuit switching simultaneously with packet switching. A new function—*hybrid ring control* (HRC)—handles circuit and frame switching.

FDDI-II uses up to 16 6.144 Mb/s channels that may be subdivided into a number of transmission channels of varying bandwidth, depending on the services to be provided. ISDN, for example, has each channel defined as an increment of 64 kb/s, making the frame size convenient for connection to the public ISDN network. The 16 channels use 98.3 Mb/s of the 100 Mb/s bandwidth. The remaining bandwidth provides a 768 kb/s packet-switched channel and 928 kb/s is used for management overhead. The bandwidth of each of the 16 FDDI-II channels is further divided into bytes of data that are spaced within 96 groups per 125-microsecond intervals. These small bytes of bandwidth ensure the predictable transmission of voice and video applications.

LOCALTALK

LocalTalk ports are standard with all Apple Macintosh computers and Apple Finder operating systems. LocalTalk's claim to fame, other than its Apple connection, is simplicity—there is almost no setup. It employs STP wiring and has a maximum data rate of 230 kb/s. Like Ethernet, it is a multiple-access method, although it employs *carrier sense, multiple access with collision avoidance* (CSMA/CA), which differs from Ethernet's CSMA/CD because its emphasis is on collision avoidance, not detection.

ARCNET

Many organizations are using or migrating to Ethernet or higher bandwidth technologies such as ATM, but there are still a considerable number of ARCNET LAN users. ARCNET began as a 2.5 Mb/s local network promoted by Datapoint in 1977. Today it reaches speeds of up to 100 Mb/s. ARCNET is organized as a logical ring and uses token-passing access. The identifications of individual terminals are physically set by an administrator and an auto-reconfiguration sequence is required if the network connections change. While it has a loyal following, its popularity is limited because Datapoint kept ARCNET proprietary, allowing Ethernet and token ring to proliferate.

Datapoint markets its ARCNET Plus, a technology that operates at 20 Mb/s, while Thomas-Conrad offers a 100 Mb/s technology. *Thomas-Conrad Network System* (TCNS), called "fast ARCNET," is a token-passing technology that retains ARCNET's MAC and memory map so that it can use standard ARCNET drivers. Perhaps even more significant to users is that TCNS can run over ARCNET's RG-62 cabling. Thus, users need not install new wire while upgrading their network performance. TCNS also runs over shielded twisted-pair, fiber-optic, and, unshielded twisted-pair cable. TCNS nodes are integrated into an ARCNET network by segmenting a distributed star topology so that only those workgroups that need high speed will incur the cost.

CONCLUSION

There is no aspect of computing that is more competitive than the local environment. LANs continue to evolve at an accelerating pace; almost before the ink dries on one standard, another rises. With the proliferation of multimedia applications, the battleground is shifting to high-speed 100 Mb/s LANs. (Witness the battle between vendor proponents of fast Ethernet.) FDDI, another 100 Mb/s LAN, has been available for several years and is evolving toward lower-cost data as well as voice connectivity. Other older LAN technologies such as ARCNET have broken the 100 Mb/s barrier and the newer 100VG-ANYLAN is emerging as a contender.

But Ethernet continues to progress, with one gigabit per second Ethernet under consideration. Several things are clear: the relatively low-cost, open connectivity that LANs offer will continue to attract users and new, powerful LAN networking technologies and topologies will enter the public telephone network in the guise of broadband WANs.

LAN Internetworking

INTRODUCTION

LAN internetworks employ a variety of networking elements. Some, like the hub, evolved from intra-LAN applications. Others, like the bridge, router, and gateway were specifically intended for internetworks. All are moving toward increasing function and performance. Industry experts foresee a time when the same technology will be used from the desktop to the LAN and throughout the enterprise WAN. As a result, the distinction between LANs and WANs will disappear because the same protocols and transmission rates will be used. Voice, data, and video will be seamlessly transported across private and public networks.

PRIVATE NETWORKS

The past decade has seen an unprecedented private corporate network expansion that has been fueled by favorable tariffs and technology. LANs have become a strategic part of the organizational infrastructure, providing distributed processing and increased user productivity. The sophistication of local networks in managing heterogeneous environments has, in many respects, exceeded that of the public telephone network. But the need for even higher data rates on LANs and the need to interconnect LANs over public networks have created a crisis in networking. Its resolution requires a radical departure from the constraints of shared-media LANs.

Networks based on emerging technologies such as ATM work differently. ATM is not a connectionless LAN protocol; it is a connection-oriented network technology. It supports hierarchical, unique station addresses that are similar to telephone numbers. These create connections between customer premise switches and carrier switches just as the public telephone network links *private branch exchanges* (PBXs) and *central office* (CO) switches. ATM will make very high-speed switched LANs feasible. When these LANs are connected to WANs, they will form the basis of the first truly universal virtual network. For many firms, considerations such as price, interoperability, and network management will be key criteria in purchasing ATM products and services. They expect ATM to provide *one* on-site interface technology that will integrate text, images, voice, and video. The public carriers have the opportunity to satisfy these needs and, in the process, bring the private networks back into the public fold.

LAN network elements such as routers and switches are also internetworking considerations. How quickly these devices evolve into true internetworking vehicles will determine whether it is the public telephone carriers or private networks that win the battle to service the broadband communications needs of modern businesses.

LAN INTERNETS

With the proliferation of LANs, it has become increasingly necessary to connect them. This may be done locally as in a single building by means of a backbone LAN. More common, though, is the use of a WAN to connect dispersed LANs to form an integrated enterprise network. The devices that remove the distance

TABLE 6.1   LAN Interconnection Devices.

	OSI layer	Device	Function
7	Application		
6	Presentation	Gateway	Connects different protocols
5	Session		
4	Transport		
3	Network	Router	Acknowledgment messages, sequence numbers and flow control
2	Data Link	Bridge	Divides a network into separate segments, ignoring the network protocol. Provides traffic balance by filtering traffic within local segments.
1	Physical	Repeater	Transfers the digital data bit-for-bit, ignoring any format or protocol. Used for increasing distance between source and destination.

limitations of LANs span the layers of the OSI reference model (Table 6.1). A LAN internetwork is formed from individual LANs that are connected by means of repeaters (OSI model layer 1), bridges (OSI model layer 2), or routers (OSI model layer 3) to a network. The network may be a backbone LAN such as FDDI, an extended LAN (referred to as a metropolitan area network, or MAN), or a WAN.

Repeaters

Repeaters are the simplest devices used to interconnect LANs. They do not control or route information, nor do they generally have management capabilities. Repeaters work at the physical layer of the OSI model and ignore higher-layer protocols while regenerating signals to extend the distance they can travel and remain recognizable to the receiving device. At the bottom (physical) OSI layer, LAN cables carry messages between sending and receiving stations as signals consisting of a series of binary digits. Repeaters connect LAN cable segments to each other. Because there is no isolation between LAN segments connected by a repeater, a single, extended LAN is created. Every bit that is heard on any one of the connected segments is repeated onto every other segment—regardless of the location of the specific device for which the traffic is intended. One machine speaks; all machines listen. Repeaters are used to extend the network because the distance a single LAN may span is limited by signal distortion. When signals become too distorted, their information is lost. Repeaters regenerate the signals and remove the distortion. In addition to extending signal distance, repeaters are also effective for linking different types of network media: thick/thin coaxial cable, copper twisted-pair, fiber, and so forth. Therefore, LANs that employ different media are often interconnected in a campus environment by means of repeaters.

There are limits to the use of repeaters (e.g., Ethernet limits the number of repeaters to four). In these cases, more intelligent network elements—gateways, routers, and bridges—are employed to extend the LAN farther.

Gateways

A gateway works at OSI layer 4 and above to convert disparate protocols and network operating systems. For example, a gateway providing access from a personal computer to an IBM mainframe would convert between the LAN and the SNA protocol environment. Gateways are often used to make E-mail systems communicate with other sites and organizations. An E-mail gateway's function is to leave messages intact by translating address headers between different mail systems. E-mail gateways that internetwork among different systems provide a common denominator between those systems, with each system supplying its own gateway to the common backbone protocol. The most common E-mail gateways are to *simple mail transfer protocol* (SMTP) and to X.400. SMTP, the UNIX/Internet E-mail protocol, is important because it is the Internet protocol. X.400, the ITU standard E-mail protocol, is important because it is the most comprehensive of all E-mail addressing protocols. Today, X.400 serves as a backbone transport solution for other types of E-mail applications. The migration from a messaging pipe to an all-encompassing messaging environment requires satisfying customer apprehensions about interoperability and directory services.

There are interoperability concerns with regard to mail systems from different vendors (e.g., Lotus and Microsoft) as well as with different X.400 versions. The X.400 standard is revisited every four years and more features are added. For example, the 1984 version of X.400 contained basic specifications for E-mail system interconnection which the 1988 version extended to encompass guidelines for distribution lists, large messages, and file attachments for various data types. This made it compatible with another OSI standard—X.500 directory services. The evolving X.500 standard outlines a way to build a common resource directory for different networks that makes it much easier to find address information or X.400 interconnected networks.

Because of the extensive amount of processing required for this level of protocol conversion, the gateway is the slowest device used to interconnect LANs. Despite the increase in computer power which speeds the processing of complex protocols, the gateway has not kept pace. Instead it has given way to faster routers and bridges that are responsible for the interconnection but relegate protocol conversions to the connecting device.

Routers

Routers are the most popular backbone internetworking device. Within the LAN, there are no dedicated paths between sending and receiving devices. To send a message, one broadcasts it to all nodes. The receiving router separates the messages (Figure 6.1). This is referred to as a *connectionless protocol*. Each device on the LAN reads the address of the message, accepts those messages that are addressed to it, and either forwards or ignores other messages. The router acts for devices on other LANs by intercepting messages and forwarding them to remote locations.

As networks grow, they are often divided into more manageable segments. When these segments are internetworked, more complex networks are formed, creating a need for alternative routes between them to provide a more reliable internetwork. Routers provide alternative paths by interconnecting networks at OSI layer 3, called the network or packet layer. A router is able to connect devices based on their logical address, which allows complex internetworks to be created. These networks can be potentially independent administrative do-

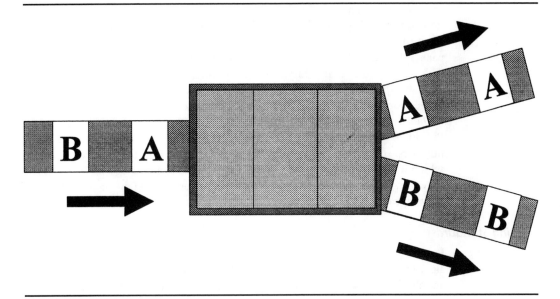

FIGURE 6.1   Message addressing schemes.

mains because the router stores a map of the entire network in order to determine the best path to a destination.

When a packet arrives at the router, it is stored until the router finishes handling the previous packet. If the packet is too large for the destination network to accept, the router segments it into several smaller packets. The router reads the destination address and looks it up in its routing table. The routing table lists the various nodes on the network, as well as the paths between the nodes and their associated costs. If there is more than one path to a particular node, the router selects the most economical path based on predetermined criteria.

The logical addresses used by the router are not unique, and naming conventions must be carefully chosen to avoid duplication. To allow this, they rely on device hierarchies in which part of the address is assigned to a group of terminals designated as either a network, a subnetwork, or an area. The remainder of the address is used to designate the particular station in the network or subnetwork.

One result of hierarchical addressing is that routers can store addressing information for networks that have very large numbers of stations. Routers also have detailed information about packet transmissions including the location of stations, packet lifetime, and the paths between nodes. They use this information to select among alternative paths by reading the frames that carry the Network layer packets of higher-level protocols, such as *transmission control protocol/internet protocol* (TCP/IP). These packets, in turn, contain the logical node numbers (e.g., IP addresses) of the sending and receiving computers. With all of this, routers can make intelligent decisions about how best to forward a packet through a LAN or WAN and ensure that it goes only where it is needed. This is important because high-level protocols like TCP/IP use broadcast frames to enable the sending computers to learn the hardware addresses of the recipient IP network nodes in order to address frames to them. Known as the *address resolution protocol* (ARP), this kind of traffic can heavily burden a large LAN. Provided with Network layer knowledge of sending and receiving IP node numbers, routers can isolate broadcast traffic to the specific LAN segments on which it is needed. This level of filtering is crucial to building a large enterprise network, which would otherwise collapse under the weight of ARP traffic if only repeaters and bridges were used.

Routing Standards

Routing algorithms are used to transport information across networks. Most algorithms automatically establish routes between network nodes from the source and destination addresses (Figure 6.2). An algorithm may employ static and dynamic routing. With static routing, the router contains a fixed routing table that contains preset routes. With dynamic routing, the routing table becomes a changeable record of a router's neighbors as well as of where each device on the network is located. Upon receiving a packet of information, the router examines the source address and compares it with entries in its table. If the source address is not in the table, it is added as a new address. The destination address is then read, compared with the addresses in the table, and also added if it is not listed.

These routing algorithms and table update methods are incorporated into the routing protocols. Various protocols exist at OSI layers 3 to 5, which exchange route and path information. Each network has its own routing protocol, which can get confusing because they route in different ways. Two of the routing protocols employed in the TCP/IP protocol suite are discussed in this section to illustrate different routing methods.

Routing Information Protocol (RIP)

This routing method, often employed in the TCP/IP-based Internet, uses the TCP/IP *gateway/gateway protocol* (GGP). The number of routers encountered by a route setup or a discovery packet as it travels from the source to destination devices determines the path. The route with the fewest routers wins. The individual routers retain routing tables that are periodically updated as each router sends a copy of its table to its neighbor. While this practice works well for smaller networks, it creates problems with larger networks for several reasons. Since routing is done without regard to significant factors such as delay and bandwidth, it is not always the best way to send information across larger

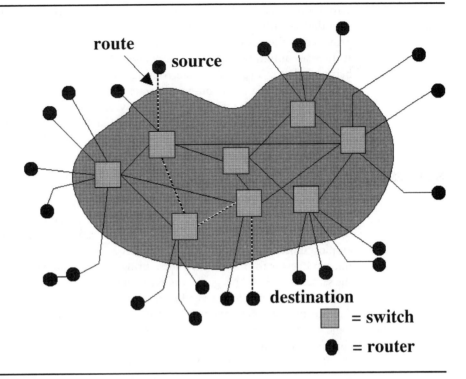

FIGURE 6.2   Router paths.

networks. As network size increases, routing updates limit the use of RIP to networks with less than 100 routers because the routing algorithm requires that the whole routing table be broadcast frequently throughout the network. Since this occurs even when there are no changes, valuable bandwidth is wasted. Moreover, when a failure occurs, this lengthy procedure slows the discovery of a new route.

Open Shortest Path First (OSPF)

OSPF addresses the RIP issues by constructing distributed routing tables. Each router on the network broadcasts a packet that describes its local links to all other routers. These description packets are relatively small and consume small amounts of bandwidth. Since OSPF issues updates only when necessary, even more bandwidth is saved. Whenever a link fails, updated information floods the network, creating new routing tables.

The table criteria include such factors as delay, bandwidth, and the dollar cost of the facility. For example, assume a packet can take either a path that goes through two routers or another path that goes through three. RIP will always take the two-router path. In contrast, OSPF chooses a path based on criteria provided by the network manager that include line speed, cost, and traffic loads. If there are multiple paths of equal weight, OSPF will balance the traffic load between them. Type-of-service routing allows the user to specify up to 16 classes of service and establish a separate path for each class. This allows the transmission of batch files over a long-delay, high-capacity satellite link and the transmission of interactive "bursty" traffic over short-delay, low-capacity terrestrial leased lines.

BRIDGES

What separates bridges from routers is raw speed. A bridge can connect at the LAN wire speed which, for 10 Mb/s Ethernet, exceeds 30,000 frames/second—a rate that only the highest speed routers can approach. (Because routers employ more complex protocols, there is more information for the router's microprocessor to process than for a bridge's microprocessor which performs the same task.) As LANs grow, the number of users may cause congestion. One way to alleviate this is to use a bridge to segment the LAN into separate parts. Architecturally, a network that is interconnected by bridging appears as a single logical network, as opposed to an inter-LAN network that uses routers and appears as a connected group of separate networks.

The bridge connects LANs at a relatively low level—the sublayer of the data-link layer. The frame is the basic unit of communication at this layer. Because bridges have access to each frame's destination hardware address (e.g., with Ethernet it is the 48-bit value that uniquely identifies each Ethernet interface card on a LAN), bridges eventually learn the LAN segment to which each device is connected. They use this information to isolate different types of network traffic according to the specific LAN segments that contain the source and destination devices. This relieves other parts of the LAN from the burden of carrying unnecessary traffic.

A bridge routes frames by means of the LLC, the upper sublayer of the data-link layer. Logically the LAN is one contiguous network, but the physical separation, if properly done, allows most traffic to remain within its segment and thereby decrease congestion. From this local application, bridges have evolved into full networking devices. In fact, some protocols such as the DEC LAT cannot be routed and must be bridged onto the internetwork. (The LAT protocol allows a terminal server to connect multiple asynchronous devices— video display terminals, printers, etc.—to a host computer.) Nonetheless, bridges remain true to their roots as data-link layer devices that forward or block traffic, depending on the source, destination, and protocol information contained in the data-link layer frame.

Whereas routers operate at OSI level 3 and connect different hierarchies of networks, bridges operate at layer 2 and connect physical station addresses within the same network. Because there is no logical separation and physically independent LANs appear as a single network, the individual LAN segments must use compatible protocols unless the bridge has a special frame translation capability such as Ethernet to token ring. Most networks that employ bridges conform to IEEE standards so that they can ensure that every device has a unique data-link layer address.

Modern bridges are able to filter information, learn device locations, and perform a rudimentary form of routing.

- Filtering: Filtering is performed by means of the frame address and control fields. The bridge reads the individual frames on one LAN subnetwork and routes between the subnetworks only those frames that have the proper addresses or control field.
- Learning: An intelligent bridge is capable of learning the locations of all devices on the LAN. The frames of IEEE LANs contain the addresses that are assigned by the IEEE to the manufacturers of LAN-attachment devices.
- Routing: Bridges employ routing algorithms such as *transparent spanning tree* and *source routing* that allow them to link LANs transparently at remote locations. Transparent bridges, using the IEEE spanning-tree algorithm in the Ethernet environment, determine a path between local and remote LANs that does not loop back to its origin. In contrast, source routing, employed in token ring bridges, sends discovery frames to determine all possible routes between local and remote LANs. Since this process also determines the network topology, closed loops can be tolerated. In order to gain the benefits of both algorithms, they were combined in the IEEE *transparent source routing* method to work with either original algorithm.

Bridge routing algorithms differ from those employed by routers in the simplicity of their interconnected network. The bridge algorithms form flat networks on which all devices are equal. In contrast, the router is able to form network hierarchies that allow routers to transport information more selectively. Routers can keep certain types of broadcast and multicast messages from entering the WAN and can route others over a single path to the destination LAN.

HUBS

Early LANs used a single, shared wire that all stations tapped into as needed. The cable was routed from office to office throughout a floor. A major problem with this approach was that a break in the cable could disable the entire LAN. The hub concept was developed to remedy this situation. Initially, hubs were essentially multiple-port repeaters. They received the signal over a wire that was terminated at a single location, preferably in a closet. Such hubs provided isolation between small segments of networked computers. If any segment failed, only the computers attached to that segment were affected. From this simple beginning, hubs have become central switches, embracing internetworking, switching, and multiple media. The flexibility of the hub accrues from its modular construction. Three basic components are used—chassis, backplane, and cards (Figure 6.3).

The chassis is the hub's most visible component. It contains several card slots and an integral power supply. The chassis usually has a high degree of redundancy in power and card connections that is achieved by means of the hub backplane design. Some chassis have separate channels that can simultaneously handle different network types (e.g., LAN or WAN). Others load-share, letting the cards select the channel that will transport the information. Still others are rigidly segmented, having specific channels for Ethernet, token ring, and FDDI, and a fast-packet bus for connecting the hub to the WAN. Whatever the current backplane design, its evolution to support higher data rates will make the hub an even more significant networking element in the future, allowing information to flow freely across network borders (Figure 6.4). The types of cards plugged into the chassis give the hub its personality. Individual cards can support different types of LANs and media, serve as internetworking devices, and provide intelligent management.

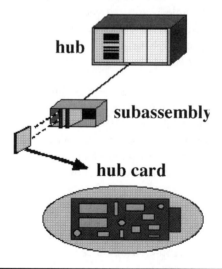

hub

subassembly

hub card

FIGURE 6.3   LAN hub construction.

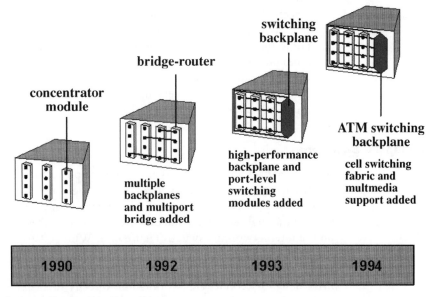

concentrator
module

bridge-router

switching
backplane

ATM switching
backplane

multiple
backplanes
and multiport
bridge added

high-performance
backplane and
port-level
switching
modules added

cell switching
fabric and
multmedia
support added

| 1990 | 1992 | 1993 | 1994 |

Source: Communications Week

FIGURE 6.4   Hub evolution.

The most important feature of the hub is its capability to alter topology. The hub's logical topology is different from the network's physical topology, so changes to the network layout—adding and deleting devices—may be accomplished without pulling cables. Wiring changes are handled by management software, a process that centralizes network maintenance. The hub concentrates a building-wide LAN's internetworking devices into a single controlled area, allowing troubleshooting from one location. From the hub, every device attached to the LAN or WAN can be controlled and analyzed. With the intelligent hub five kinds of switching are possible: port, bank, card, Ethernet, and cell.

- Port Switching: An administrator can assign any given port in a card to any logical network.
- Bank Switching: The same as port switching, with the addition that any given port can be made part of a logical grouping of ports by the administrator.
- Card Switching: The network administrator can switch a bridge, router, or repeater card from one backplane bus to another to balance the traffic.
- Ethernet Switching: Individual Ethernet frames are switched based on the destination address. This treats the LAN as a PBX rather than as a party line.

Ethernet switching bypasses the CSMA/CD protocol used by Ethernet to deliver the full 10 Mb/s bandwidths to each port up to the bandwidth capacity of the hub. There are two switching designs: One sends the frame out to the destination port before it has arrived at the input port; the other relies on wire speed control of each frame.

- Cell Switching: A variant is ATM in which the data is segmented into 53-byte cells. The smaller cells may be switched at higher speeds than the much larger frames that they replace.

Collapsed Backbones

A collapsed backbone hub offers an even more powerful integrated framework for managing enterprise internetworks. Although such hubs will continue to provide the physical connectivity for workgroups and departments, a hub port will serve a single computer and hub traffic will be concentrated to a single port on the internetworking switch (Figure 6.5). With collapsed backbones, problem isolation will be simplified because when failures occur, it will be obvious which computer failed. These sophisticated hubs will fundamentally change LAN design by connecting workstations directly to switches and providing a new level of visibility into network traffic. The switch will have router functionality because the higher-layer network management required will, typically, be in a centralized computer room. In Figure 6.5, Ethernet traffic previously used for inter-workgroup or interdepartmental traffic is collapsed onto the backplane of a centrally located internetworking switch. Fiber runs vertically up on risers joining with hubs on each floor of the building. The hubs connect with Ethernet and token ring LAN terminals by means of unshielded twisted-pair wiring. The fiber backbone supports the heavy traffic as well as injecting fault tolerance into the building internetwork.

It is fortunate that everything communicated on a LAN passes through the hub in the wiring closet. A smart hub can understand the hardware and IP addresses at the beginning of each message, learn everyone's location, and handle all routing requests from its central location. Combining the features of repeaters, bridges, routers and switches, the collapsed backbone hub could begin transmitting an output frame on the proper segment when it reads enough bits from the input frame, acting like a LAN bridge that is almost as fast as a repeater. It can also perform more sophisticated tasks such as using the IP addresses in the packet to learn where all users reside and handle ARP requests for them, without sacrificing plug-and-play simplicity.

SWITCHES

Intelligent hubs are being challenged by more powerful enterprise switches. A LAN switch works by adding a high-speed backbone (that is, a switching fabric)

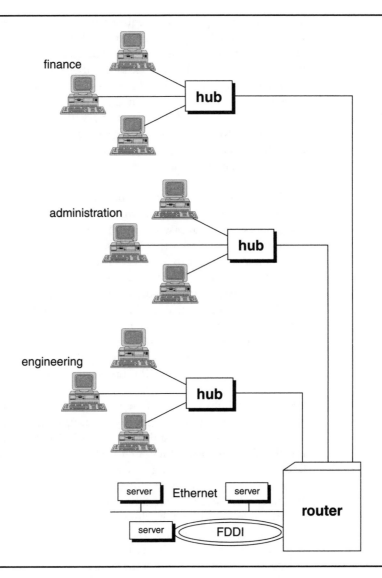

FIGURE 6.5　　　Collapsed Backbone Network.

such as ATM to connect multiple LAN elements to each other. Each switch port then provides a dedicated, point-to-point link for the user. Most organizations use twisted-pair Ethernet (10Base-T) to interconnect PCs and workstations through a common repeater. The 10Base-T wiring, which is the bulk of the physical plant, uses ordinary telephone wire and terminates in telephone jacks that can be patched as required when users move from office to office. Everyone on the LAN shares Ethernet's 10 Mb/s bandwidth. This appears to be enough bandwidth; compared to a wide area network T-1 (1.544 Mb/s) it is, but it can become a bottleneck as the user community grows and new networked applications are brought on-line. By replacing the repeaters with LAN switches, however, everyone in the workgroup gains a dedicated 10 Mb/s network path. For a LAN with 50 users, this increases users' available bandwidth to 500 Mb/s without having to change office wiring or interface cards. More benefits become apparent as the limitations of other types of LAN Extenders—repeaters, bridges, and routers—are examined.

- Repeaters: Operating at the physical bit-level layer of the OSI reference model, repeaters cannot optimize LAN traffic.
- Bridges: There are two kinds of frames in a LAN: directed frames, hardware addressed to a specific interface card; and broadcast (or multicast) frames, addressed to a logical group of interface cards. Bridges are smart enough to isolate directed frames, but they must forward each broadcast frame they receive in all directions. There is not enough information at the data-link layer to enable a bridge to filter this kind of traffic.
- Routers: Unlike Ethernet repeaters and bridges, routers are difficult to configure. One has to assign network and subnetwork numbers—a finite resource—to each router's ports. For this reason, it would be impractical to install and configure an IP router for each user in any large network.

Like a LAN hub, the switch segments LANs into more manageable parts, providing efficiency and economy. Today, commercially available switches now support between 4 and 90 ports with data rates from 45 to 155 Mb/s. Initially, the more powerful switches had traditional asynchronous WAN interfaces and supported existing as well as more advanced network connections (Figure 6.6). Now, as newer switching fabrics such as ATM spread throughout WANs and LANs, most interfaces will use ATM. The massively parallel architecture of an ATM switch allows concurrent switching along many parallel paths, giving each port full use of the allocated bandwidth. Employing cell switching, these switches break up data streams into very small units that are independently routed through the switch.

Another benefit, along with cell switching and scaleable switching fabrics, is network management. Spanning the boundaries between LANs and WANs, the enterprise switch will offer sophisticated management that reduces the problems of changes and moves. Workgroup users can be registered together, regardless of their physical location. The switch automatically reconfigures the list of addresses when a registered network user plugs into a port, even when the new port is on a different LAN.

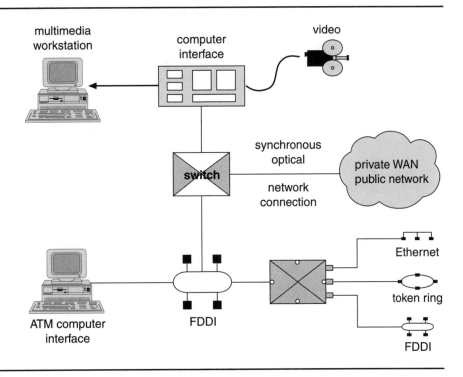

FIGURE 6.6   Enterprise network switch.

CONCLUSION

Although repeaters, bridges, gateways, and routers have evolved separately, they are merging into the LAN hub with the additional function of switching. The hub, in turn, will be eclipsed by the enterprise switch. On the basis of concurrent rather than shared media technology, such switches allocate full bandwidth to the user for the required interval. The importance of the enterprise switch will increase as multimedia applications proliferate, more terminals are connected to hosts by means of LANs, more PCs are networked through LANs, and minicomputers, workstations, and mainframes are designed to use LANs as the primary host-to-host and host-to-front-end access method.

Since all of these data sources have placed new traffic burdens on the WANs to which they connect, a new hierarchy in WANs has been developed that allows LANs to extend beyond their traditional boundaries. An extended LAN, the range of which may encompass a city or a metropolis, is referred to as a MAN. Considerable work is still required to flesh out its standards, particularly in the area of management. As first-generation LANs are phased out over the rest of this decade, most current management problems will be resolved. Fiber LANs should become the preference for high-capacity private networks. SONET will facilitate the interconnection of far-flung MANs. MAN traffic at the T 1, T 2, and T 3 rates will be able to access the SONET backbone via virtual tributaries. Meanwhile, data traffic at the FDDI or fast Ethernet rate of 100 Mb/s will be mapped onto a SONET/SDH OC-3 payload for transport to a remote MAN. At the same time, WANs will rely less on circuit switching, as embodied in TDM-based multiplexers and more on high-speed packet technologies.

CHAPTER
7

MANs to WANs

INTRODUCTION

The survival of businesses depends on access to the right information at the right time, which in turn allows the mobilization of proper responses to customers, competition, and opportunities. The current trend toward corporate downsizing and moving staff closer to the customer has resulted in global enterprise dispersion. This phenomenon has created the need to internetwork millions of local computer environments. Most U.S. companies are actively engaged in interconnecting LANs. Even more are concerned with communications between dissimilar mainframes, minicomputers, and other office automation vehicles. Broadband networking is being driven by many forces; principal among them is the increasing need among LAN users for high-speed data services. Two trends are responsible for multimegabit data communications:

1. The need to interconnect traditional LANs over an even higher-capacity backbone LAN.
2. The dramatic increase in computing power and the equally dramatic decline in its cost.

One result of these trends is a bewildering—and continuously developing—array of new carrier services that seamlessly internetwork LANs and WANs, satisfying the demand for sophisticated data, image, and video communications (Table 7.1). Emerging services include *switched multimegabit data service* (SMDS) offered by the Telco/PTTs and BISDN/ATM offered by LAN equipment vendors and Telco/PTTs. All are dependent upon newly emerging technologies. SONET/SDH forms the basis for linking high-speed LANs, while ATM realizes the full potential of bandwidth-on-demand services. Together they integrate

TABLE 7.1   Advanced network services projection.

Service	1993 share of total revenues	1998 share of total revenues	2003 share of total revenues
frame relay	94%	38%	1%
SMDS	5%	37%	4%
ATM	1%	24%	60%
broadband ISDN	—	1%	35%
total	$696 million	$26 billion	$138 billion

Source: Electronicast Corp., San Matco, California

LAN and WAN network elements, using common equipment and signaling, and making possible a rich variety of network architectures.

This section describes the types of broadband networks presently being implemented. These networks are not growing in a vacuum for emerging applications continue to stimulate the demand for broadband equipment and services. Consider, for example, the value of having hospitals globally linked. Patients could have access to the best doctors and diagnostic routines no matter where they are located. A doctor in Los Angeles could use a software program to discover and explore anomalies in a patient located in Moscow. Magnetic resonance images could be evaluated in real-time by a supercomputer and transmitted anywhere they are needed, eliminating hours, even days, of delay.

Broadband network growth is being encouraged by the popularity of distributed networks whose resources are either shared by many users or are dedicated to specific users as their needs may warrant. The demand among businesses for distributed processing and high-speed networking, unconstrained by geographical limitations can only be fulfilled by the emerging broadband network hierarchy. But instead of connecting from the LAN into the broadband WAN directly, a new form of intermediary is being deployed—the MAN (Figure 7.1). Its purpose is twofold: first, to provide a regional conduit for broadband traffic; and second, to gather the volumes of information that warrant an economical connection to a broadband WAN. Of the many proprietary and standard forms of the MAN, the most popular is SMDS.

SWITCHED MULTIMEGABIT DATA SERVICE (SMDS)

SMDS is a specification for a metropolitan area, high-speed data service distributed by the *Regional Bell Operating Companies* (RBOCs), some *Interexchange Carriers* (IXCs), and several European public network providers. SMDS supports a

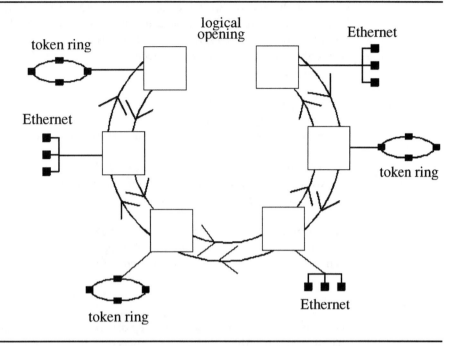

FIGURE 7.1   Metropolitan Area Network (MAN).

variety of high bandwidth applications, providing the economies of shared access within metropolitan areas. It allows the same type of connectivity for data that the telephone offers for voice and fax. Information is transported over counter-rotating rings that extend over a municipality. Its users gain switched access to a network that provides the following:

- A high-speed data service within a metropolitan area
- Features similar to those found in LANs, in particular, high bandwidth and low delay
- Easy integration with existing systems and the capacity to evolve gracefully with them
- Security features, such as closed user groups
- A connectionless, high-speed packet service

Because SMDS is a metropolitan-area network specification, it has been limited to use within *local access and transport areas* (LATAs). The inter-area connections needed for a national network have been harder to come by. Although coast-to-coast communication among local SMDS networks is possible through standard direct lines, local networks have not employed consistent implementations of SMDS. Consequently, it has been a harder sell than the providers anticipated. Despite this, SMDS now has hundreds of customers and is emerging as a niche service that meets the communications needs of companies with the following basic profile:

- Multiple, geographically dispersed locations, each with its own LAN and/ or host computer system.
- A need to exchange information among their own locations or with other businesses.
- An expected growth in data traffic that will require a high-speed backbone with greater capacity than is currently available or affordable.
- A preference for a public network rather than a private one.

Such organizations benefit from the ability of SMDS to interconnect distributed computing applications. Because SMDS supports both existing and emerging technologies, it provides the growth needed for future applications.

Technology

SMDS is a connectionless, cell-switched data transport service that combines a shared-medium LAN with ATM and offers switched access. It was conceived as a high-speed public packet service, mainly to interconnect local area networks. The SMDS standard defines the following tiered architecture:

1. A switching infrastructure comprising SMDS-compatible switches (which may or may not be cell based).
2. A delivery system made up of T1 and T3 circuits called *subscriber network interfaces* (SNIs).
3. An access control system for network managers to connect to the switching infrastructure without having to become a part of it.

As with a LAN, SMDS requires an access technique to prevent overlapping transmissions. It is also like a LAN in that it is connectionless—it does not set up the sequence that has become known as a virtual circuit. Each packet or datagram is addressed and switched independently, with no prior network connection. SMDS differs from LANs that use either CSMA/CD or token-passing techniques because it uses a distributed reservation scheme in which each node keeps count of the access requests made by the nodes ahead of it.

SMDS utilizes the IEEE 802.6 protocol, which transmits ATM cells over a shared bus. The 802.6 protocol supports a complete mix of services: constant bit rate, variable bit rate, and available bit rate. Data packets are transferred across an SMDS network, employing a *distributed queue dual bus* (DQDB). DQDB is the

IEEE 802.6 cell relay network standard for switching and transmitting 53-byte cells. With DQDB there are two buses that pass data in opposite directions (Figure 7.2). Stations transmit on both buses in opposite directions, after queuing their data and waiting for the headend station of the bus to grant permission to send. This station continuously transmits empty cells that travel onto and down the bus and may be used by the other stations to transmit information. When the empty cells reach the end of the bus, they are discarded. The last station or tail station monitors the bus and will disable devices that are monopolizing the bus. The buses are usually configured in a loop for greater reliability. When a failure such as a cable break occurs, the two adjacent stations temporarily serve as upstream and downstream masters.

Each SMDS packet has the capacity for as many as 9,188 bytes of data, which accommodates entire packets from most legacy LANs—Ethernet and token ring—except for 16 Mb/s token ring that can have a maximum frame size of 16,000 bytes. The SMDS packets are divided into cells or slots of fixed length—53 bytes per cell, of which 48 contain data and 5 are for control. The lower layers of *SMDS interface protocol* (SIP) transmit the 53-byte cells in conformance with the ATM structure. Because 802.6 was developed early in the ATM standardization process, it uses AAL 3/4 cell format instead of the AAL 5; therefore, it requires an 8-bit access control field rather than the 4-bit generic flow control field of conventional ATM. Otherwise the cells, by design, are ATM cells. Thus, DQDB delivers a compatible format to carry SMDS data and

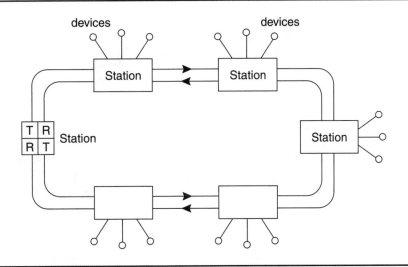

FIGURE 7.2   DQDB looped bus.

interface with ATM networks that are accessed across a dedicated link, typically from a customer's network. According to current specifications, access is by means of a DS1 or DS3 line, with service classes defined for a sustained transfer rate of 4, 10, 16, 25, and 34 Mb/s. A SONET/SDH interface at STS-3/STM-1 is expected in the future.

SMDS Internetworks

The LAN-like features of SMDS make it a natural backbone network for seamlessly interconnecting Ethernet, token ring, FDDI, and ATM LANs over extended geographical areas. In a MAN configuration, SMDS can interconnect users for distances up to 50 kilometers. Beyond that, interconnection by means of a SONET-WAN is desirable, with DS3-based networking as an alternative (Figure 7.3). To connect a LAN to an SMDS network requires only a router and an SMDS-compatible DSU/CSU or SMDS host adapter card. For users to communicate over the WAN from one SMDS/MAN to another requires a gateway or bridging function. Interface guidelines, developed by the SMDS Interest Group, support the networking protocols found in LAN environments—TCP/IP, Novell's IPX, AppleTalk, DECnet, SNA, and OSI. The SMDS Interest Group, a consortium of internetworking and wide-area network vendors, also designed an SMDS interface (Figure 7.4). The interface allows LAN routers to connect to SMDS networks via data service unit/chan-

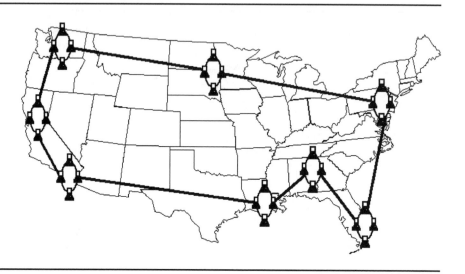

FIGURE 7.3   MAN internet.

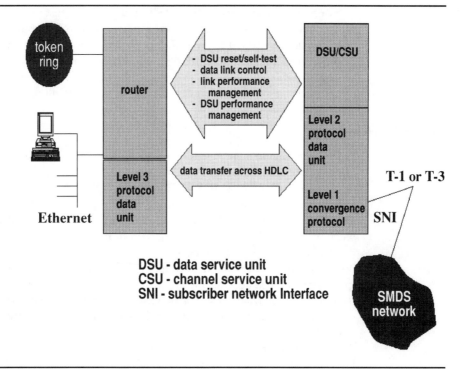

FIGURE 7.4   Proposed SMDS interface.

nel service units, or DSU/CSUs, and to partition SMDS functions between the router and the DSU/CSU. The interface is based on a *high-level data-link control* (HLDC) protocol that is widely used for communications between LAN and DSU/CSU equipment. The bridge/router may map the DS1/DS3 signal to SONET VT1.5/STS-1 prior to a SONET NE interface. The SONET NE would provide DS1/DS3/OC-1/OC-3 extensions as required.

SMDS does not require carrier switches to establish a call path between two points of data transmission. Instead, SMDS access devices pass 53-byte cells to a carrier switch. The switch reads addresses and forwards cells over any available path to the desired end point. SMDS addresses ensure that the cells arrive in the right order. One benefit of this connectionless service is that it puts an end to the need for precise traffic flow predictions and connections between fixed locations. With no requirement for a predefined path between devices, data can travel over the least congested routes in an SMDS network. In fact, SMDS cells are released into the public network delivery system as letters are to a post office. By knowing an SMDS address, subscribers can call up and send/receive data.

Because SMDS coexists with dedicated facilities, customers can create hybrid public/private networks. Existing networks can be easily expanded and new sites quickly added to an SMDS network without totally reconfiguring it. Additions to an SMDS network only require an update to a screening database on the SMDS switch. Since SMDS is connectionless, its users may construct mesh networks within which each site is connected to all other sites.

Carrier Services

SMDS data services, at speeds from 56 Kb/s to 45 Mb/s, are offered in the United States and Europe. (The *European Technical Standards Institute* (ETSI) has developed a version of SMDS for European public networks. ETSI adapted the U.S. version of the standard to work over European E1 (2.048 Mb/s) and E3 (34 Mb/s) data trunks. ETSI also made a few modifications that allow SMDS to work with the signaling and framing used in European leased lines.)

Initially conceived as a switched-access service at T1 and T3 rates, SMDS's popularity increased when the service was offered at lower access rates. Slower speed service became available when several of the companies that make the *channel service unit/data service unit* (CSU/DSU) equipment (which connects customers' equipment to the local exchange) collaborated to provide a lower-rate interface. This digital exchange interface encapsulates cell-based SMDS datagrams into variable-length frame packets that can be carried over 56 Kb/s lines and converted back into SMDS 53-byte cells at the other end. CSU/DSUs for 56 Kb/s connections are significantly less expensive than those for higher-speed T1 or T3 services. The routers and other equipment that support the digital exchange interface are compatible, which is not always the case with specialized equipment that uses the higher-speed SMDS interface protocol.

The effect of the digital exchange interface standard has been to make SMDS competitive with frame relay. The cost of subscribing to the service is about the same as for leasing several frame relay ports. (SMDS data services are roughly $200 per month for 56 Kb/s connections and $800 per month for T1.) Nonetheless, in order to become a viable service, SMDS will have to offer higher bandwidth than frame relay does and be able to seamlessly interconnect with other carrier services, neither of which is yet possible.

Interconnection services are offered by independent telephone companies and IXCs. Co-Net Communications, Inc. of Orlando, Florida, an SMDS network operator that is aimed at the graphics industry and MCI, among others, offer interLATA connections. The pricing of these services depends on customer equipment, software, store-and-forward capabilities, and interLATA connections that are based on the total transmitted number of megabytes. To illustrate, MCI's HyperStream SMDS prices are comprised of two components: a monthly port/ subscription charge based on access speed and transport usage charges based on the number of megabytes delivered. Transport usage charges are set with a minimum and maximum range by access port speed per location. MCI prices do not include applicable LEC SMDS service charges. SMDS is more economical than private line when the network contains at least four fully meshed sites. The more sites and the greater the distance, the greater the savings that are derived from using MCI HyperStream SMDS. In a typical network configuration, an SMDS solution would cost less than half that of a five-node fully meshed private line network with a 500-mile average distance between locations.

Business Applications

The needs for MAN services, and for SMDS in particular, are quite varied. Any data transfer requiring large amounts of bandwidth, such as the interconnection of large numbers of subscribers to a wide-area public data network, imaging, computer-aided design, publishing, and financial applications, can benefit from SMDS. The following sections describe several actual applications.

Publishing

Avanti Press, Inc., a Miami-based company that prints department store catalogs and corporate annual reports, uses SMDS technology for data exchange between its two facilities in Miami and New York and its subsidiary in Rochester, NY. Initially, the company needed to transmit graphics files as large as 500 megabytes, which were overloading its modems and could not be hand-carried on floppy disks. Other applications followed. At its Case-Hoyt facility, Avanti prints corporate annual reports and also designs, creates, and prints catalogs for large department stores including Saks Fifth Avenue and Bloomingdales. During the proofing stages, SMDS allows Avanti and its clients to transfer accounting data and graphical images back and forth. An average file size is about 75 megabytes, but it can be as large as 500 megabytes and as small as 12 megabytes. The clients' equipment acts as a receiving station for the data, through fully configured computers temporarily set up by Avanti. No operator is needed on the client end to complete the file transfer. The cost of hooking and unhooking SMDS is low enough to warrant its use even for short projects that last only two to three weeks.

Other publishers agree. Pre-press house Group Infotech of East Lansing, Michigan, depends on SMDS to communicate with publishers in New York and printers in the South, as well as to handle local jobs over Ameritech Corp.'s LATA SMDS network for sophisticated color work and OCR.

Internet Access

The *Commercial Internet Exchange Association* (CIX) adopted SMDS as the technology for high-speed connections between its members. The CIX is a trade association whose members operate the largest commercial Internet exchange point, located in Santa Clara, California.

Interexchange carriers MCI Communications Corp. and Packets, Inc. are vying to help companies gain commercial Internet access through switched multimegabit data services or SMDS. Like the Internet, SMDS has unpredictable traffic flow among sites. The need for high-speed connectivity on an occasional basis has made SMDS attractive to a number of publishing companies. Widely dispersed groups of artists, contractors, pre-press bureaus, printers, and publishers with large data-transfer needs can communicate through SMDS within a LATA and even outside a telephone company's service area.

Banking

Meridian Bancorp, Inc.—a financial corporation that consists of a banking company with 300 branches, a mortgage company, a securities company, and an asset management trust company—uses SMDS as a backbone for interconnecting its headquarters, data center, mortgage company headquarters, and several regional offices at T1 rates. Savings are considerable compared with the cost of dedicated lines. Initially, Meridian used 56 kb/s dedicated lines to connect the data center and the headquarters of eight different companies. The system handled such high-speed applications as LAN interconnection, IBM token ring traffic, mainframe access, and connections to a DEC VAX system. As the demands on the network increased, it became apparent that there was too much traffic on the 56 kb/s lines and not enough bandwidth, especially for large file transfers and for the multiple bank users who were constantly accessing the firm's commercial loan database for real-time updates.

Meridian wanted to upgrade to a higher-speed data transport system that would also provide the flexibility it needed to meet changing requirements and the ability to ensure cost-effective expansion. Finding a solution with these capabilities was essential since Meridian's management wanted to be able to interconnect the Fort Lauderdale securities location with the Pennsylvania main offices, and eventually interconnect all of its separate bank branches. Meridian now uses T1 SMDS as a wide-area network backbone to link 11 locations including its headquarters, data center, mortgage company headquarters, and several regional offices. The company also has an SMDS line set aside for network-wide backup for vital financial transaction data. This SMDS solution cost only one-third of that for a 56 Kb/s private line solution and two-thirds for a T1 solution.

Government

The State of Delaware moved its data communications off a point-to-point multidrop T-1 network onto a statewide SMDS (Figure 7.5). Delaware's relatively small geographic size makes it ideal for SMDS because the entire state is served by only one LATA and the lack of interLATA SMDS is not a problem. Nearly 300 SMDS lines are connected to the state's SMDS network, transport-

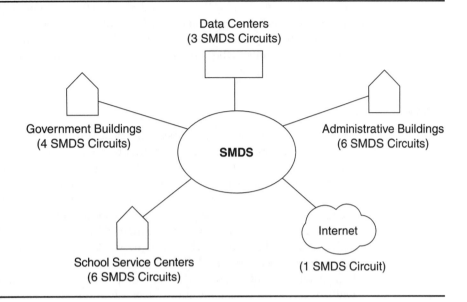

FIGURE 7.5   State SMDS network.

ing two main network protocols: Banyan Vines Internet protocol routing and TCP/IP. State employees use workstations and PCs to access the network for a wide variety of applications that include electronic mail, asynchronous database distribution, mainframe access, client/server applications, and Internet access.

TRANSPARENT LAN SERVICE (TLS)

Offered by US West, Ameritech, and Teleport, among others, LAN Interconnect Service uses a very high-speed, shared medium that employs a time-division bus to provide an SMDS-like service. This is a brute force, bulk bandwidth service that transports information in circuit-switched channels which are similar to those of a multiplexer inasmuch as subscribers share the medium and the service is not tariffed; therefore, the rates are lower than SMDS. Using US West's pricing as a gauge, TLS and SMDS have comparable costs for 3-node systems with a 10-megabit rate. Although TLS has mileage charges and SMDS does not, the expense for adding more sites increases roughly fourfold with SMDS. The more locations that participate, the more the use of a shared medium helps.

With TLS, the telephone company installs a standard LAN interface, generally Ethernet or token ring, on the customer's doorstep. This is connected to the premise network. The TLS box sends the data to corresponding boxes at the customer's other sites, appearing as a LAN bridge to the user. This service uses proprietary trunk protocols between sites and does not offer the data security of SMDS. Its customers who are companies with large amounts of data in local geographic areas, rent entire TLS rings from the provider.

WANs

There are many varieties of WANs. The most common is the global public-switched telephone network. As a data transport, the public telephone network is losing strength because it was originally designed for voice traffic and only recently has seen the balance of traffic shifted to data. Under the assault of the data applications that are emanating from LANs, asynchronous network elements are failing to meet the need for higher speed and better managed information highways. Many types of data networks (e.g., X.25 packet) are also global in scope, but it is rare that a data network or private corporate network does not interface with the PSTN. The need to coexist with the PSTN has influenced the architecture of data networks, constraining the amount of bandwidth available for data applications. The solution is an intelligent, broadband public telephone network that uses SONET/SDH transport and ATM switching to create faster and more powerful information superhighways (Figure 7.6).

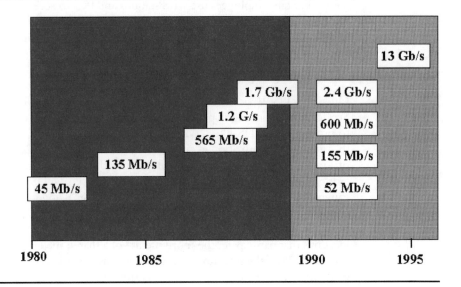

FIGURE 7.6   Increasing network speeds.

Public Telephone Internetworks

When information is transported from one site to another, it often traverses the public telephone networks of different carriers, both local and interexchange. No matter which network is used, the average bandwidth-to-switch voice call is 64 Kb/s or DS0—the bandwidth increment used by voice switches. The telephone network uses an asynchronous signal hierarchy that has a DS0 bandwidth granularity. Even with lower rate lines at the carrier voice switch, a DS0 is used (Figure 7.7). A channel or DS0 supports one or more voice conversations, depending on the compression algorithm that is used. For example, PCM supports one voice conversation; *adaptive PCM* (ADPCM) supports two conversations; and proprietary compression algorithms are for 3 to 16 conversations while maintaining toll-quality voice. Since a customer pays the network provider for bandwidth regardless of the amount of information that is crammed down the lines, it is advantageous to use compression. Although voice channels can be multiplexed with data, voice must have priority because voice traffic cannot tolerate delay.

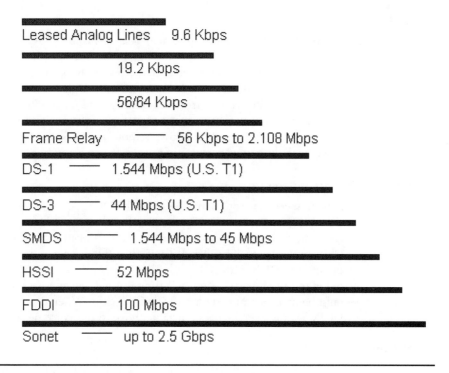

Leased Analog Lines   9.6 Kbps

19.2 Kbps

56/64 Kbps

Frame Relay   ——   56 Kbps to 2.108 Mbps

DS-1   ——   1.544 Mbps (U.S. T1)

DS-3   ——   44 Mbps (U.S. T1)

SMDS   ——   1.544 Mbps to 45 Mbps

HSSI   ——   52 Mbps

FDDI   ——   100 Mbps

Sonet   ——   up to 2.5 Gbps

FIGURE 7.7   Line speed.

If computing needs were to remain static, the narrowband WAN infra-structure—and consequently the disparity between LANs and WANs—would probably exist for years to come. However, the ways in which people interact with computers are changing, requiring more bandwidth. For example, users are likely to run on-line background applications that receive updated information in response to complex database queries. More and more applications use multimedia and incorporate audio, images, and video. These applications demand more bandwidth from the WAN than is readily available without broadband transmission rates of 45 Mb/s and greater.

Although data networks are logically separate entities, they may use the same facilities and physical cabling as the PSTN does. In order to transport high-bandwidth applications, data networks employ their own broadband network elements for routing purposes.

Broadband WAN Elements

The network elements described in this section form the basic building blocks for networks, whether public or private.

CSU/DSU

The purpose of the *channel service unit/data service unit* (CSU/DSU) is to encapsulate information in the proper framing before it enters the WAN. The CSU/DSU provides the interface between the on-site hub, switch, or router formats and the broadband network. It regenerates the signals received from the network and can also serve to troubleshoot the transmission line. The CSU/DSU automatically monitors the signal to detect violations and signal loss. When problems are discovered, it allows remote network testing from the central office, including loop-back testing of the transmission line.

Digital Cross-Connect Systems

Higher-speed data routes may be set by the DCS. Another function of the DCS is *groom and fill*, which entails the selective removal of DS0s from a T1 facility (grooming) for routing to designated remote locations and the placement of DS0s onto a partially utilized T1 facility (filling) for efficient transport. Although the DCS is more flexible than a manual patch panel, it is not as well suited to real-time tasks such as setting up calls and alternate routing because it must be preprogrammed with the desired paths. When data is transported with DCSs, it runs over a separate data network to the voice network (Figure 7.8).

Multiplexer

The slowest defined data rate for broadband networks is faster than that at which most routers, bridges, and video *CODer/DEcoders* (Codecs) operate.

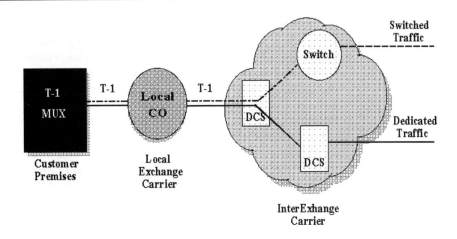

FIGURE 7.8   DCS routing of switched and nonswitched traffic.

Even at T1 rates, a *time division multiplexer* (TDM) is used to share the bandwidth. It switches among the 24 channels, interleaving the bits into one continuous digital stream of 1.544 Mb/s. Multiplexing is done in pairs. At the source, a multiplexer interleaves the bits from various sources, while at the destination a multiplexer separates the bit stream into the 24 channels.

At higher data rates, multiplexing becomes even more important. The network manager has the choice of dedicating a high-capacity circuit to each low-speed device (possibly wasting bandwidth) or multiplexing several lower-speed signals onto a broadband circuit. A broadband multiplexer may be considered similar to a switch, but it does more than just provide connectivity to multiple users; it handles isochronous data such as voice and video. The multiplexer accepts data from routers and bridges through a standard interface (V.35 or HSSI). Then it segments the data into cells, addresses each cell, and maps the cells into a WAN framing structure (T3/E3 or SONET/SDH).

Router

The router provides protocol conversion between LAN packets and broadband wide-area networks. For example, a router can convert between LAN frames and ATM cells transmitted over an ATM-switched network. A SAR function within such a router will first implement an AAL to convert LAN frames into cells for transmission, then will reassemble received ATM cells into LAN frames. Eventually this capability will allow ATM to be used for the high-speed backbone transmission of LAN and WAN traffic, while individual users continue to maintain their existing LAN standards and cost structure.

Switch

The public network switch is a larger and more intelligent version of the enterprise switch. A public switch is capable of handling hundreds of thousands of cells per second and has thousands of switch ports, each operating at SONET/SDH rates (Figure 7.9). All cell-processing functions are performed by the input controllers and the switching fabric. In a broadband switch, cell arrivals are not scheduled. The switch takes all incoming packets, divides them into 53-byte cells, adds headers, then retransmits them onto the network. If SONET/SDH is used, the available bandwidth has already been divided into slots (Figure 7.10). The control processor resolves contentions when they occur, as well as performing call setup and teardown, and bandwidth reservation, maintenance, and management.

The *input/output* (I/O) ports are synchronized so that all cells arrive at the switch fabric with their headers aligned. The resulting traffic is said to be *slotted* and the time to transmit a cell across the switch fabric is called a *timeslot*. All VCIs are translated in the input controllers. Each incoming VCI is funneled into the proper output port as defined in a routing table. At the I/O ports, the cells are arranged in the proper transmission format. For example, a broadband ISDN port provides a line terminator to handle the physical level transmission and an exchange terminator for cell processing.

FRAME-RELAY NETWORKS

Communications systems that hold the line open, empty and waiting for the next burst of information, force users to pay for more bandwidth than they need. Packet-switching networks allow a much more economical transmission of computer data. The X.25 protocol commonly used in packet-switched networks transports a standard amount of data packaged with an address. When a data packet arrives at a node, the node checks for errors and if it finds none, reads the packet's address and routes it accordingly. Packets arrive at their destination and are assembled by the receiver. The benefit of X.25 packet-switching is that messages for many different destinations can share the same transmission facilities at any given time.

Frame relay is this generation's technology alternative to older, slower transport networks such as X.25 and *synchronous data-link control* (SDLC). It is predicted that users could increase throughput by 30 to 50 percent by running SNA traffic over frame relay rather than SDLC. A public frame-relay network is a multinode network of switches with dedicated connections to customer sites (Figure 7.11). Frame relay is often the access protocol into the network switch, although different protocols may be employed between the switches. The service provider permits access to its switch by means of DS0 and T1 lines. These leased lines have a fixed speed that is negotiated with the carrier. As these customer-to-carrier *point-of-presence* (POP) lines are relatively short, their rental

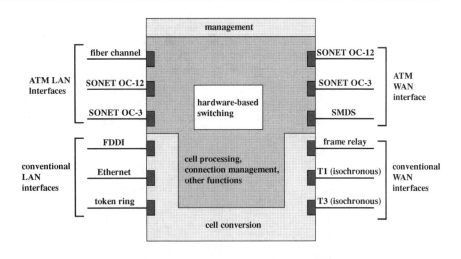

FIGURE 7.9   Broadband switch.

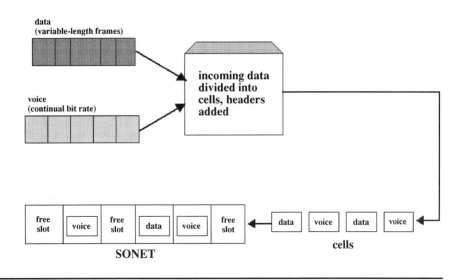

FIGURE 7.10   Broadband switching.

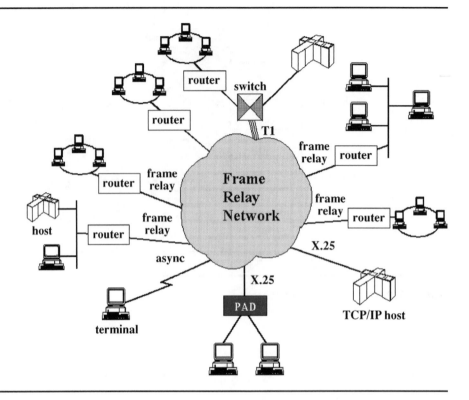

FIGURE 7.11   Frame relay network.

expenses are small and a customer's traffic is aggregated with that of other customers to save long-haul transport costs.

The attractiveness of frame-relay services to users is their cost effectiveness. The services can interface with existing customer hardware at a minimum upgrade price and may be optimized for transporting the bursts of traffic that are characteristic of LAN internets. Customers are charged for the number of PVCs and the connected end-points. The PVC data rate is the *committed information rate* (CIR), which is the maximum average data rate of each PVC. This rate can be exceeded during bursts of LAN traffic, an advantage to users.

Frame-relay networks provide:

- Efficient accommodation of bursty applications that require high-speed data transport.
- Cost-effective alternatives to leased lines, especially for full logical mesh connectivity.

- ANSI and ITU–approved, standards-based infrastructure.
- Support for the interconnection of LANs running different protocols.

The need to define PVCs between devices is well suited to SDLC whose users are accustomed to dedicated, nonswitched data paths. But PVC-oriented frame relay is not optimal for the connectionless LAN environment in which the use of SVC connections are preferred because such circuits provide network resources only when necessary. Carriers and vendors have worked together to resolve this issue. In 1996, frame-relay carriers began offering SVC for frame relay. Unlike PVCs, SVCs provide connections between two points in the frame relay network on an as needed basis. The route through the network is established only for the duration of the transmission, then it is dropped. Paths through the network do not need to be preprogrammed, only the source and destination addresses. This makes it easier to configure frame-relay service and significantly expands the market. Instead of limiting access to a predetermined number of sites, SVC enables connections to any location with an SVC address.

Frame-Relay Service

Frame relay is an established WAN service that offers guaranteed high bandwidth at moderate prices. As a data transport it fills the gap between X.25, ISDN, and T1 leased lines. Telephone companies that supply frame-relay services base their rates on a CIR. These carriers price frame-relay lines cheaper and more flexibly than leased private lines. Some businesses save as much as 50 percent on telecommunications costs by switching from low-speed leased lines to frame relay. For example, in 1995 AT&T charged 18 percent less for frame-relay service of similar bandwidth than for a private T1 line from Des Moines, Iowa to San Francisco, California. Wiltel Communications Corp. charged 41 percent less for a 64 Kb/s frame-relay meshed network between Chicago, New York, San Francisco, and Los Angeles than if it leased a 56 Kb/s private line covering the same area.

Frame-Relay Applications

Cost savings are enticing many private-line customers to use frame-relay services that are accessed from their premises by means of a router that supports frame relay or a *frame-relay access device* (FRAD). Frame relay service is attractive for metropolitan and regional applications, especially where there are many sites in a compact area, as in a school district, municipal government, or bank. Actual examples include:

- The Los Angeles County Office of Education is using frame relay to connect 50 sites throughout the county.

- Several municipal governments have switched from dial-up modem networks to frame relay for cost and performance reasons.
- In setting up its network of 10 branch offices, Cambridge Savings Bank, in Cambridge, Massachusetts, connected to frame-relay services. The bank uses frame relay as its primary WAN link and has ISDN in place as a backup service.

To date, however, the applications that frame relay can support are limited. Lower-speed applications can be satisfied, but most higher-speed ones will have to wait for faster networks such as ATM. Frame relay's data rates do not support voice or video. These will continue to be relegated to guaranteed-fixed-rate T1/ T3 or will need to be handled by faster broadband networks. When these faster networks become available, frame relay either will be assigned to slower traffic (much as X.25 is today) or will serve as the access protocol into the faster networks. In order to relieve users' anxieties that frame relay might only be an interim step on the way to ATM, most carriers intend to provide interoperability between frame relay and ATM.

ATM-BASED NETWORKS

The remote LAN internetworking needs of businesses continue to grow, offering an excellent opportunity for ATM-switching networks.

LANs

LAN bandwidth steadily increases. Today FDDI and fast Ethernet can push the bandwidth of LANs to 100 Mb/s. However, upgrading shared-media LANs simply delays the bandwidth congestion problems that are inherent in all shared LANs. The capacity of shared-media networks is only as great as the speed of the common bus and extra bandwidth cannot be added. In contrast, ATM-based LANs allow users to connect to the network at their required speed, adding bandwidth as needed. ATM removes the limitations of shared LAN backbones by collapsing them into the ATM switching fabric.

The typical extended ATM LAN will employ an ATM switch to link slower LANs. To interconnect a series of LANs, multiple switches can be concatenated together (Figure 7.12). The switches will be more powerful than today's intelligent LAN hubs in terms of processing power, routing capability, and network management. The ATM switch is further distinguished from today's hubs in that its total bandwidth is the sum of the bandwidths of its input ports. With an ATM switch, information can travel from any port to any other port without being blocked, in contrast to traffic on bus-based hubs that can be blocked if there are too many sources on the bus.

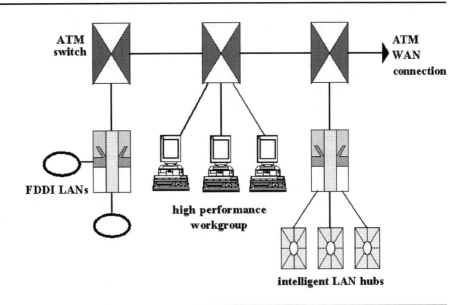

ATM
switch

ATM
WAN
connection

FDDI LANs

high performance
workgroup

intelligent LAN hubs

FIGURE 7.12   Concatenated ATM switches.

Public Broadband Networks

Future broadband public networks will blend cell-relay switching and synchronous optical transmission, creating an ultrahigh frequency communications fabric. A single multiservice platform will support POTS, ISDN, frame relay, SMDS, and BISDN. These networks will transfer huge files—gigabytes heading toward terabytes—from one site to another. Information will flash across the globe at the speed of light. The benefits will be astounding; hitherto insoluble problems in computing will be solved. The networks will be used for metacomputing in which remote mainframe computers are tightly coupled for real-time cooperative processing. Users will dynamically allocate bandwidth among a variety of services and redirect traffic between subnetworks. Voice communications, video exchanges, and supercomputer data will all be simultaneously switched as conveniently as data alone is today.

The future begins with today's ATM switching services involving MANs, frame relay, and LAN interconnection. The competition is fierce as new players—IXCs, RBOCs, alternative access providers, and cable TV companies— offer ATM services. The services, which are currently being marketed on a contract basis, are divided into four categories: Class A, a constant-bit-rate service, is similar to private-line connections for voice and data traffic; Class B involves video; Class C is a variable-bit-rate arrangement similar to frame relay; and Class D is a connectionless service comparable with SMDS.

Telco/PTT

Full-ATM based, wide-area networks from the public carriers began rolling out in 1994. Today, long-distance carriers, alternative carriers, and regional Bell operating companies have nearly 100 switch sites up and running. Among the RBOCs:

- Pacific Bell became the first Bell company to provide ATM service when it introduced a 155-Mb/s offering in San Francisco. It has since added 1.5- and 622-Mb/s services and expanded its coverage.
- Ameritech offers ATM in Chicago, Detroit, Indianapolis, Milwaukee, Cleveland, and Columbus with Dayton to follow. Users are able to feed frame-relay traffic to the carrier's ATM backbone, run both local-area and wide-area ATM backbones, and even take ATM to the desktop.
- Bell South and Bell Atlantic introduced ATM service for specific government customers. North Carolina's state government is the beneficiary of the Bell South offering, while Bell Atlantic's service is to the Department of Defense and several other federal agencies in the Washington, DC area. Its ATM tariff covers constant- and variable-bit rate service at 45 and 155 Mb/s.

ATM switching services are available from several long-distance companies (Table 7.2). (It is also offered by international PTTs in France, Germany, and

TABLE 7.2   U.S. ATM Network Providers

Provider	Headquarters
AT&T	Basking Ridge, NJ
Ameritech	Hoffman Estates, IL
Bellsouth Corp.	Atlanta, GA
GTE Corp.	Stamford, CT
MCI Communications Corp.	Washington, DC
MFS Datanet Inc.	San Jose, CA
Pacific Bell	San Francisco, CA
Sprint Corp.	Kansas City, MO
Teleport Communications Group	Staten Island, NY
US West Inc.	Englewood, CO
Wiltel Inc. now LDDS	Tulsa, OK

Japan, among others.) Sprint is constructing an ATM network for the U.S. Department of Energy and NASA. This network operated at 45 Mb/s in 1993, 155 Mb/s in 1994, and 622 Mb/s in 1996. Sprint intends to add BISDN and SONET capabilities and will combine its private-line and circuit-switching voice services with ATM-based services under the ATM backbone. GTE SPANet provides a real-time gigabit-rate fusion of voice, video, and imagery.

Other Providers

The Federal Communications Commission's decision to allow competitive access providers to collocate their equipment in local exchange carriers' central offices has encouraged ATM networks from new competitors. Among the independent carriers, MFS Datanet, Inc. has built a nationwide fiber-optic network supported by ATM switches that extends from the United States to overseas with networks in London, Paris, and Frankfurt. The network provides customers with *high-speed LAN interconnection service* (HLI) at native LAN speeds. HLI service includes long-distance transport, associated hardware, and software support, and 24 hour monitoring and maintenance with a 90-minute response time for service. Science Applications International Corp. (SAIC), a systems integration company, plans a less ambitious 45 Mb/s transmission service. SAIC will use Wiltel Communications Systems' fiber-optic network and ATM switches to provide a nationwide switched broadband service that will deliver high bandwidth for interLATA interconnection. Their network could be used to interconnect frame relay or SMDS, or it could transmit combinations of multimedia data.

BISDN

BISDN is a service platform. Using a limited set of network interfaces and network equipment configurations, a single BISDN network can support a wide range of customer data, voice, and video applications. BISDN takes advantage of the availability of high-capacity synchronous optical fiber networks (SONET/SDH) to reliably transport enormous quantities of information. Access to BISDN is at 155 Mb/s and 622 Mb/s using high-speed, fixed-length ATM cells. With that much capacity, BISDN is expected to carry a multitude of services.

With BISDN, private networks require only a small number of physical interfaces. Users needing 45 Mb/s for high-quality videoconferencing have only to dial up a 45 Mb/s connection. Virtually any bandwidth between network locations can be established on demand. The list of services that could be provided with BISDN is just beginning to unfold. They fall into two basic categories: interactive and distributed. Interactive services include videoconferencing, video telephony, high-speed data, electronic mail with images, and interactive database services with high-resolution imaging and audio enhancements. Distributed services include cable TV (existing quality), high-definition cable TV, pay-per-view TV, and compound document distribution.

INTEGRATED SERVICE DIGITAL NETWORK (ISDN)

After decades of languishing in the backwaters of telephony, ISDN has become a reality. The generic software of most telephone switches will support out-of-band ISDN signaling, making ISDN service generally available. (In 1995, 80 percent of domestic households had access to ISDN on their telephone lines.) The RBOC service is gradually gaining subscribers, overcoming the earlier stigma of too much promise and too little delivery. Nonetheless, there are still barriers to a wider adoption of the technology. Many ISDN products are difficult to set up and be made to work with other equipment because they are poorly installed and lack proper diagnostics. Setting up line parameters with the telephone company is not only difficult, but is worsening as more devices compete for the ISDN line.

Standards-based compression is another technology needed to help ISDN gain "critical mass." Many vendors use proprietary compression schemes, but they are of limited usefulness because most telephone company switches only support 56 Kb/s per channel. A final obstacle is the state *Public Utilities Commission* (PUC) regulation. Some commissions understand the importance of intelligent ISDN tariffs, but many continue to lag and others, such as in California, are using strategies that in effect raise rates.

Despite these impediments, ISDN services fill an important gap in the information technology managers' repertoire. Because ISDN can provide bandwidth on demand, it can support disaster recovery scenarios. It has the capability for simultaneous voice and data transmissions on a single line and advanced digital network features such as *automatic number identification* (ANI). Most importantly, ISDN overlays the existing asynchronous public telephone network with digital signaling, thereby providing more reliable data service.

Technology

The ISDN extends the digital fabric of the public-switched telephone network the last mile to the subscriber premise. There are two versions of ISDN in use today: a basic rate interface that employs a two-wire 1.44 kb/s full-duplex connection from the telephone company's central office and a primary rate interface that delivers T1 service over four wires.

The *basic rate interface* (BRI), also known as 2B+D, consists of two 64 kb/s B channels (DS0s) and one 16 kb/s D channel for common signaling and control (Figure 7.13). Signals move full duplex across the U interface at 80 kb/s. Basic rate signals are transported on a 160 kb/s data stream that consists of 144 kb/s of bearer and delta channels, 12 kb/s of framing, and the remaining 4 kb/s for a maintenance channel. Some devices attach to the U interface. Another interface point, the S/T interface, is a four-wire, multipoint interface that allows several devices to share one ISDN circuit. The network terminator, NT1, converts data between

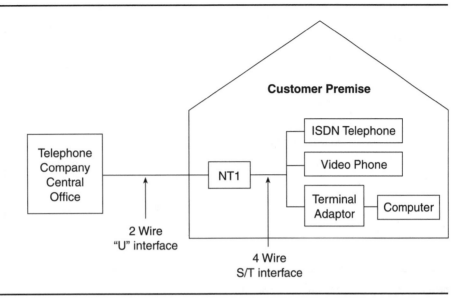

Customer Premise

Telephone
Company
Central
Office

NT1

ISDN Telephone

Video Phone

Terminal
Adaptor

Computer

2 Wire
"U" interface

4 Wire
S/T interface

FIGURE 7.13   ISDN basic rate interface.

the U and S/T interfaces. The NT1 passes the ISDN data stream back and forth between the local devices and the two-wire telephone lines. Each B channel can carry data or voice conversations at the same time that the D channel transports signaling such as call setup and call progress information.

When a telephone call is destined for an ISDN-to-ISDN connection, the digital telephone network passes the encoded data stream to the customer premises. A call setup packet is transmitted that includes the destination address, identifies the calling party, and contains information on how the call was routed. If conventional connections are made for fax or modems over an ISDN circuit, the signaling is ignored, making ISDN compatible with the existing base of analog circuits and equipment.

The primary rate ISDN PRI or 23B+D carries, depending on the country, 23 or 30 64 kb/s DS0 channels and one 64 kb/s D channel for common signaling and control. It employs the same signaling protocol as the base rate. Primary rate is used to terminate data calls from basic rate subscribers, to serve as a trunk to PBXs, or to provide dial-up T1 lines for disaster recovery purposes.

Application

A growing number of users are turning to ISDN because it provides digital service to their desktops and eliminates modems. ISDN also employs a switched

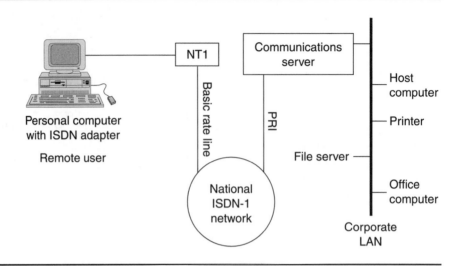

FIGURE 7.14   ISDN network application.

line rather than a dedicated line, making it an economical supply of reserve bandwidth because customers do not have to contend with the 24-hours-a-day charges of dedicated lines. ISDN service subscribers only pay for what they use.

ISDN is at its strongest in remote site and branch office internetworking where transmission speeds of between 56 Mb/s and 256 Mb/s are required. An increasingly popular application, telecommuting, allows a remote user to place a 64 kb/s data call to a communications server in order to access a corporate database (Figure 7.14). The communications server supports the internetworking of PCs on ISDN interfaces within the corporate network. The server is set up with a multiline hunt group that allows users to dial into the same number, simplifying administration and spreading access ports uniformly among available lines. The information transfer can occur at the full speed or be rate-adapted to compensate for slower equipment in the transmission path.

NISDN

In 1991, the slow deployment of ISDN caused a group of regional Bell operating companies, switch vendors, computer manufacturers, and large end users to commit to a revised Bellcore-driven technical standard known as National ISDN-1. The National ISDN-1 package is less comprehensive than the original ISDN Phase 1.1 Technical Reference. It represents a compromise between specifications developed by Bellcore and the North American ISDN Users' Forum. The purpose of the *narrowband integrated services digital network* (NISDN) is to accelerate the distribution of ISDN services. NISDN in some regions is subdivided into basic rate interface and primary rate interface. For NISDN, the SONET RFT may be more appropriate than an ADM. Its BRI ISDN interface would provide two DS0s for the 2B channels (64 kb/s), while the 16 kb/s D channel for signaling and control could share a full-time DS0 with three other D channels in a 4:1 time division multiplexing arrangement. The primary rate service of 23B+D or 24 DS-0s would be a full DS-1 interface mapping into floating VT1.5s of STS-1.

CABLE TV NETWORKS

Cable television systems began as a way to direct television, radio, and data signals from a central originating location to residences and businesses by means of coaxial cable. As a result of consumer demand for more reliable, interactive services, this is changing. Cable TV networks in the United States are now being transformed into multiservice broadband networks. In order to understand the impact of this on the cable TV provider, it is useful to trace the evolution of the cable TV network.

Cable TV systems were devised to access television signals in poor reception areas by receiving them in a central area called the *headend* where the signal is amplified and rebroadcast over coaxial cable in trunk lines to feeder lines to cable drops to individual homes. Signal fidelity is maintained by amplifiers that are located at quarter-mile intervals along the network routes. By 1992 cable TV companies employing this type of unidirectional network had grown so they had access to 97 percent of all U.S. households and connected 61 percent of them. This ubiquitous presence is only exceeded by that of the telephone industry.

Cable Distribution Plant

The distribution plant consists of three main elements:

1. Headend: In a traditional cable network, the headend is the source of the signals received by the subscriber. It is equivalent to the telephone company CO and is the point at which all the program sources are received, assembled, and processed for distribution (Figure 7.15). A headend can range from a small utility building to an elaborate control center consisting of thousands of pieces of equipment, including satellite, radio, and television receivers, and towers, antennas, computers, TV production studios, and test instruments.

2. Feeder Plant: This is the physical coaxial cable that either is stretched across power or telephone company poles or is buried underground to connect the headend to the residence or business community area. It consists of a trunk system and feeder network. The trunk uses large diameter cables

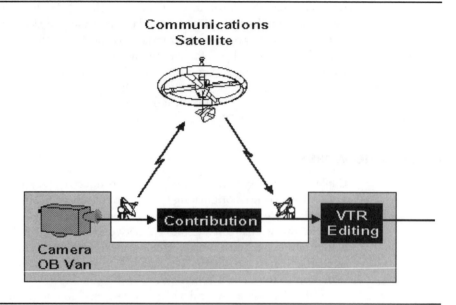

FIGURE 7.15   Broadcast chain production.

(three-quarters to one inch diameter) between the headend or hub and the community, that split at various drop points. Trunk cable serves to transfer signals to the subscriber community. As the electronic signals flow along the trunk, they lose energy, so amplifiers are periodically placed along the path to boost the signal to an acceptable level. The feeder system is the network that parallels each street within an area and is the point of connection for subscribers.

3. Subscriber Drop: This refers to the small diameter (about one quarter of an inch) coaxial cable that leads from the street into the individual subscriber's home or office as well as to the equipment that connects to the subscriber's television.

Microwave and satellite are sources of signals to the cable TV headend as well as receivers for signals from the headend. Cable television systems are limited in area to a radius of about five miles from the headend location. To service large metropolitan areas, a single headend is insufficient and a hubbing arrangement is employed. The area around the main headend is served by trunk lines going from the headend. Other areas of the community are served by remote hubs, each of which handles an area within a five-mile radius. Most cable systems are now interconnected to allow common programming and advertising. Nearby systems may be interconnected in the same ways that hubs are linked, but systems that are too far apart must rely on microwave or satellite signals.

At present, cable TV networks are optimized for the one-way delivery of a common menu of programs to all of the homes on a network. For cable operators to participate in the emerging interactive services market, their networks must be upgraded to supply customers with real-time, two-way communications, improved reliability, and support for individualized programming.

Traditional Network Deployment

The way in which the network is deployed depends heavily on its existing structure, the area it serves, the population density, and the installation cost. There are many alternative architectures for routing between nodes. The feeder plant is the most crucial for determining network capabilities in telephony, multimedia, and video-on-demand. Traditional cable TV networks use coaxial cable as the feeder plant media, stringing one inch thick trunk cables out from the headend for distances of 10 to 15 miles. Every 2,000 feet, an electronic amplifier is placed to regenerate the signals carried along the trunk cable.

The physical topology of a Cable TV feeder network is a bus. Any signal bound for a location on the network passes by all other network ports along the way. Usually, a common set of outbound signals is broadcast to every port on the network. Inbound signals flowing from a single port toward the headend pass by all the upstream ports on the segment of distribution cable that serves the residential neighborhood. Larger trunk cables carry signals from the headend into neighborhoods. Part of the signal is redirected or tapped off for local distribution within an area that might contain several hundred homes. The demarcation point between the feeder network and the subscriber drop is a bridge amplifier that may be located within a trunk amplifier housing.

Hybrid Networks

Perhaps the most significant improvement in trunking is the advent of optical fiber (1,310 nanometer and 1,550 nanometer optical wavelengths). A hybrid distribution system uses a star physical topology based on a combination of optical fiber, coaxial cable trunks, and copper-dedicated links between the headend and small serving areas (Figure 7.16). There is a direct, dedicated, and optically passive connection between the headend and an optical receiver to the subscriber's neighborhood. The connection between fiber and copper is accomplished in broadcast applications with a receiver node. The node converts the optical signal into an electrical signal. Future nodes will be capable of converting the digital optical signal into an analog electrical signal and addition-

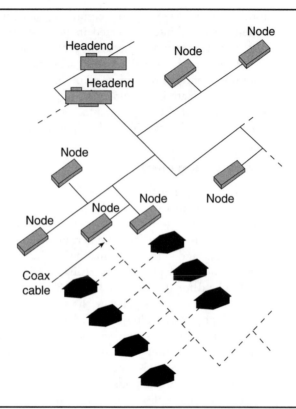

FIGURE 7.16 Hybrid fiber coaxial network.

ally handling and converting the upstream feedback from the subscribers. The expectation is that every node will serve 80 to 500 homes to provide telephony services as well as traditional entertainment programming.

The telephony interface unit connects the headend equipment with a local central office via T1 connections. The headend interface unit appears to the telephone company switch as a DLC. At the headend, the digital telephony signals are demultiplexed and modulated to an RF digital carrier for transmission over the cable system. The telephone signals to be carried over the cable system are terminated at the cable drop and passed through the residence as a standard two-wire telephone conversation. The customer interface is located outside the subscriber's home. It terminates the broadband cable signal, converts it to analog form, and distributes it through existing wiring to provide POTS and cable service.

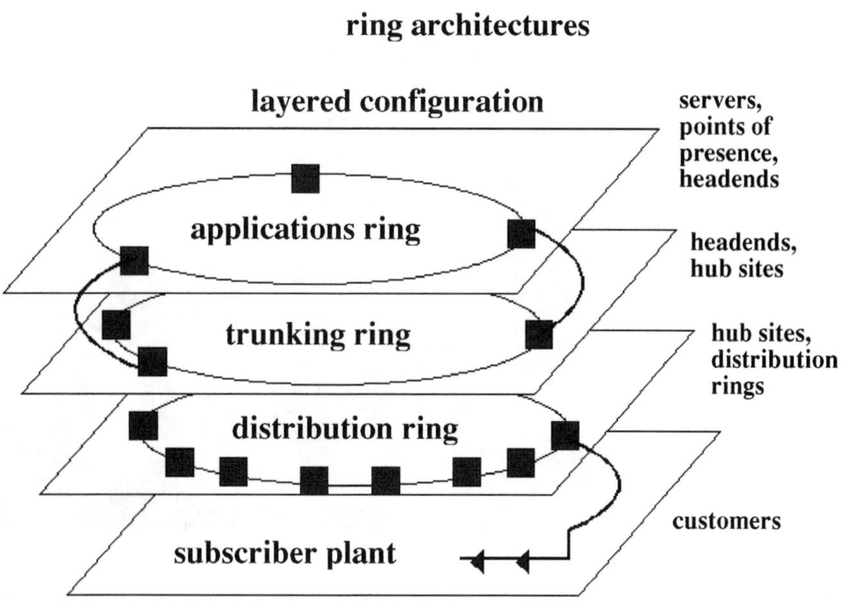

ring architectures

layered configuration

servers,
points of
presence,
headends

applications ring

headends,
hub sites

trunking ring

hub sites,
distribution
rings

distribution ring

subscriber plant

customers

FIGURE 7.17   Cable TV networks.

Cable TV Network Rings

Today's cable TV networks are being built on optical fiber platforms and incorporate the network redundancy available with SONET rings (Figure 7.17). This "ring of rings" structure features signal redundancy at three levels: headend-to-headend, headend-to-hub, and hub-to-feeder.

- Applications Ring: The applications ring connects headends, local-exchange carrier central offices, long-distance telephone carrier POPs, and video-on-demand and multimedia playback centers (Figure 7.18). It resembles the ring networks that link a local exchange carrier's serving offices with tandem offices and interexchange carriers. These rings operate at OC-3 or higher.
- Trunking Ring: Whenever a single headend location cannot serve all customers, a trunk ring is used to extend access distances to about 11 or 12 miles. The hubs on the trunk ring can be simple signal-repeater locations fed by digital or amplitude modulated signals over fiber or they can be microwave-receiving stations. Sometimes, hubs can add local off-air broadcast or satellite signals to the signal stream.

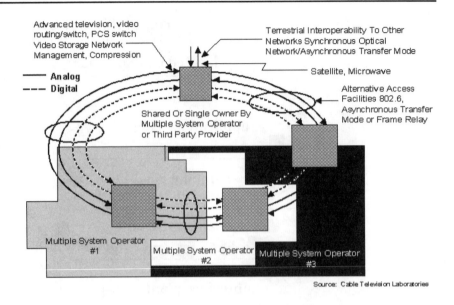

FIGURE 7.18   Regional network architecture.

- Distribution Ring: The distribution ring carries telemetry data and up-stream signals for interactive programming and services in the 5 to 30 MHz spectrum. The 54 to 550 MHz band can support 75 analog video channels. Bandwidth from 550 to 850 MHz can carry 500 channels of compressed digital video, and the 450 MHz between 850 MHz and 1.3 GHz can be reserved for telecommunications and personal communications services.

The ring topology based on SONET will support a variety of transmission protocols including ATM, FDDI, and SMDS. The deployment of this architecture will be driven by the development of the markets for new services (e.g., video-on-demand and personal communications) as well as by competition between the cable TV companies and LECs for basic telephone service.

CONCLUSION

The success of broadband Telco/PTT services remains somewhat uncertain especially for the local loop. Here private lines are the most popular corporate data communication vehicle with the lines typically affording speeds from 56 Kb/s to 45 Mb/s. The stakes for MANs are high—nothing less than the competitive health of the local telephone company in data services. Unfortunately, the telephone company MAN service, SMDS, is priced too high for some users. As for WANs, it appears that ATM and frame-relay services are supplanting private lines; while ISDN, an innovative technology, has yet to find a receptive user base.

The information systems manager has a decision to make about deploying broadband networks. Clearly all the processing power of today's computers is relatively ineffective without efficient interconnection. Broadband networks are able to work with existing networks as backbones for private networks and as overlays for public networks. Broadband services are readily available. The selection between them is predicated upon a clear understanding of what they can and cannot do, and how they fit into existing corporate networks. Unfortunately, until standardized rates and tariffs are widely published, most managers will be forced to make ad hoc decisions.

BROADBAND NETWORK TECHNOLOGIES

CHAPTER

8

Broadband Frame and Cell Technologies

INTRODUCTION

Every application has an optimally sized data packet or frame for transporting information over a network. Real-time applications favor smaller packets since they are less likely to create bottlenecks. Most existing data networks favor larger packets so that they can move bigger chunks of data. With the X.25 slow-packet standard, some public carriers settled on a 128-byte packet, because this allowed the best network throughput when errors and retransmissions were considered. New packet-transmission technologies are now being deployed that achieve orders of magnitude greater than the throughput rates of X.25. One such technology—fast-packet—is already a success in private T1 networks. Fast-packet is a generic term that is applied to many different high-speed transmission technologies (Figure 8.1). It achieves high performance by eliminating many of the overhead functions that are carried out at intervening nodes by X.25. Fast-packet networks exhibit low latency and very high-speed switching and thus are suitable for broadband communications. Frame relay is a fast-packet standard capable of transporting both small and large size packets at speeds up to T3.

Frame Relay

The frame-relay protocol is a data transfer protocol defined by the ANSI and the *International Telecommunications Union* (ITU). It is similar to the X.25 and ISDN standards, but it assumes a reliable transmission medium and therefore contains very little error-recovery functionality. As a result, it is more straightforward and the data transfer is more efficient. At present it is used mainly in North America and Japan. A typical application is LAN-to-LAN interconnection.

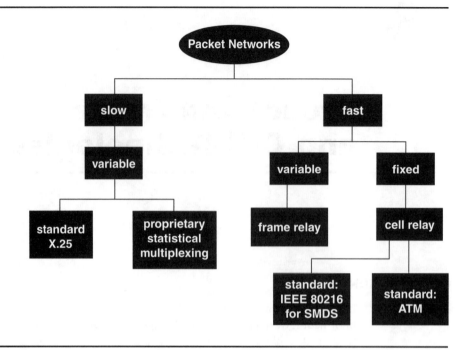

FIGURE 8.1   Types of packet networks.

Frame relay operates at OSI model layers 1 and 2 and resembles LAN networking protocols. It assumes that the customer is using error-correcting protocol suites such as TCP/IP, IBM's SNA, Digital's DECnet, and Novell's IPX/SPX, among others, as well as relatively noiseless fiber and digital transmission. In contrast, X.25, developed for older analog leased lines, operates at OSI layers 1 through 3 and provides extensive error control. While frame relay resembles a stripped down X.25, it combines reduced overhead with high-speed interconnecting trunks and operates at 64 kb/s rates and above. Overhead is reduced by eliminating the error checking and correction done in the data link (layer 2 of the OSI reference model) and by moving the function of network layer 3 to increasingly intelligent end-stations (Figure 8.2). X.25 networks perform extensive error checking and correction because they were designed to compensate for poor quality analog communication lines and dumb end-stations. For a packet network to work in that climate, it had to verify the integrity of the information that it was transporting.

A primary difference among packet technologies is the way in which the address and control field bits are used (Figure 8.3). By extending its address field,

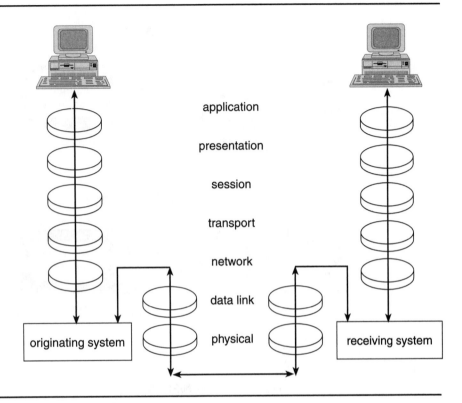

FIGURE 8.2   OSI end-terminal interconnection.

frame relay performs basic frame routing and control within the data-link layer. An abbreviated destination address is processed at each node by a frame handler that has a list of routes between end-stations. There are three key elements in the address field; the *data link connection identifier* (DLCI), the *forward and backward explicit congestion notification* (FECN/BECN) bits, and the *discard eligibility bit* (DE).

- The DLCI is instrumental in routing frames from one node to another by identifying a frame's logical channel within a shared physical line. It defines a virtual circuit and its type as *switched virtual circuit* (SVC), *permanent virtual circuit* (PVC), or *multicast virtual circuit* (MVC). A DLCI terminates at the receiving port of a node and may be regenerated with a different value at the transmitting port as the frame travels node-to-node between two end-stations (Figure 8.4). Between nodes, a physical line can simultaneously support up to 1,024 logical channels or connections. Depending on the

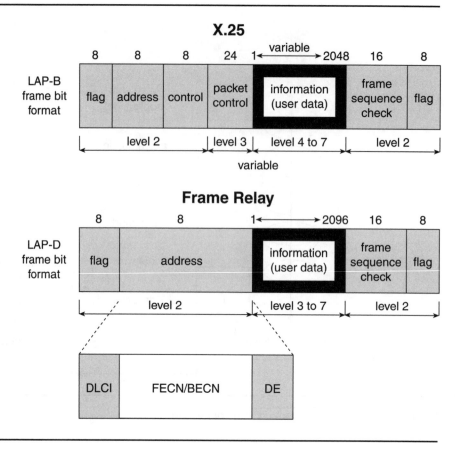

FIGURE 8.3   WAN frame formats.

network architecture, the number of nodes, and the structure of the routing tables at each node, a frame traveling from one end-station to another may use several DLCI values before it reaches its destination.

- The BECN/FECN bits (and/or the DE bit) determine the network response to congestion. When congestion occurs, a frame-relay network may control the traffic flow either at the source, at the destination, or at both. The BECN bit controls traffic at the source, while the FECN bit initiates flow control at the destination.

- The DE bit sets the priority for frames to be discarded during periods of congestion so that lower priority frames are discarded before higher priority ones.

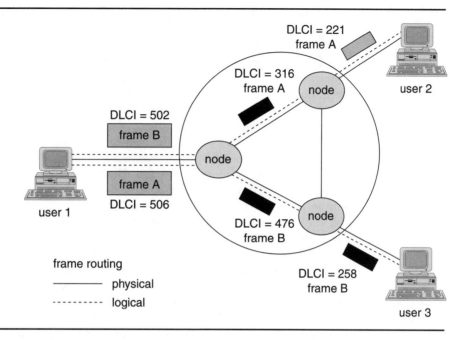

FIGURE 8.4   Use of DLCI for transmission.

Cell Relay

Cell relay employs tiny frames to transport information at even higher rates than frame relay. At the time of transmission, the cells are filled by data packets that arrive at irregular intervals (asynchronously). The line of cells itself is synchronous, transporting a continuous stream of data. But packets enter the stream only when they are available. The process resembles the loading of an endlessly circulating train of box cars. Each car in the train travels in synchrony with the others, but the rate at which information is loaded or removed is asynchronous. When there is a packet to be shipped, it is dropped into a car. At any one time, packets from different sources may contend for a specific box car, but only one packet can be loaded at a time. The rest must wait until a car is available, forming a queue. This process introduces a variable delay across the train. Sometimes there are many cars in a row, all filled with data packets. At other times only one car is filled, leaving a large gap between it and the next car that contains a packet. Since an observer looking at the train in the middle of its run would not notice a reproducible pattern in the rate that the filled box cars pass, the cars or cells are deemed to be transferred asynchronously. The maximum rate that information travels is the rate at which the train is traveling.

When a user wants to send information to another user, the originating user signals the receiver with the cells. The first transmitted cell in the message contains the number of box cars of information. This cell goes to the destination. There it is removed and the destination sends a cell back to the source by a returning car. This cell defines the rate at which information may be transferred. For example, one box car full of information must be followed by five empty cars. This process, called signaling, is used to establish a connection and to negotiate throughput, grade of service, delays, and so on.

Cell relay uses small packets for high-speed, low-latency transport. Large amounts of information can be transmitted without monopolizing the network for long because the information is segmented. The cells are formed from the contents of much larger data words and are then rapidly switched and interleaved together. Software programming is too slow for controlling this process, so hardware is used with silicon integrated circuits as the enabling technology. Cell segmentation and reassembly equipment would be too expensive without employing specialized silicon integrated circuits that are specifically designed for the task.

Several emerging broadband carrier services are based upon fixed- and variable-sized cells. Fixed-sized cells have several advantages over the variable-sized ones used in carrier services such as SMDS. Fixed-sized cells reduce the queuing delay for high-priority calls because they are more rapidly processed at the user network interface. They are also switched more efficiently, which is the key to achieving the very high data rates achieved by the ATM form of cell relay.

ASYNCHRONOUS TRANSFER MODE (ATM)

Telephone companies (Telcos) and their overseas equivalent *private telephone and telegraph companies* (PTTs) are migrating to standard network technologies to support the high bandwidth data communications needs of their business customers. Frame relay provides high-speed access, but lacks the capacity for a backbone Telco/PTT network. SONET/SDH networks have the bandwidth capacity, but lack the switching fabric. ATM fills this void. It was designed as a telecommunications technology, specifically as the transport for Broadband ISDN (BISDN). These standards define two ways to transfer blocks of information across a network—a *synchronous transfer mode* (STM) and an asynchronous transfer mode (ATM).

- STM, which should not be confused with SDH's STM-1, is a time division multiplexing method used in digital voice networks. It allocates time slots every 125 microseconds within a synchronously recurring frame. Multiplexing and switching equipment divide the total network bandwidth into a hierarchy of fixed-size channels such as DS0, DS1, DS2, DS3, and so forth. Each STM channel is identified by the position of its time slot(s) within the 125-microsecond frame.
- ATM allocates the total network bandwidth to services such as data, video, or

voice transport and flexibly shares both bandwidth and time. ATM offers switching at a variety of speeds because it is not associated with a single medium or transmission speed. Unlike STM, which breaks down bandwidth into channels of information, ATM transmits fixed-length packets of information or cells whenever a service requires bandwidth. It transports continuous bit-rate traffic such as voice and video as well as noncontinuous traffic such as the bursty data frequently encountered with LANs.

Despite the fact that it was developed for public telephone networks, ATM found its first application in LANs. As long as LAN users continue to transfer ever-larger data files and attempt to integrate isochronous voice traffic with data traffic, there is a great need for ATM. Nonetheless, its widespread use in LANs could still benefit the Telco/PTT by creating a demand for ATM services. Carrying LAN traffic over WANs is a challenge because LANs employ different technologies and have different transport needs from WANs. LANs rely on communication standards that support high data rates over relatively short distances. WANs use communications standards that work over longer distances, but are relatively slow. Should WAN traffic become ATM-based and the same technology be used in the LAN, carriers will face a considerably easier task of integrating WAN and LAN traffic.

Broadband Switching

The ATM switching fabric is data-rate independent and supports both public network and LAN switching at ultrahigh rates—exceeding 1 Gb/s. All ATM switching uses standard 53-byte cells. Each ATM cell has a 5-byte header that contains virtual circuit and virtual path identifiers (Figure 8.5).

The header is transmitted first and contains the addressing information. It does not carry any service-specific data. Rather, it defines the user/network interface by means of the following fields:

- *Generic flow control* (GFC)—the 4-bit GFC is used for end-to-end flow control.
- *Virtual channel identifier* (VCI)—the VCI (16 bits) is similar to an X.25 virtual circuit; it defines a local logical connection between two ATM nodes.
- *Virtual path identifier* (VPI)—the VPI (8 bits) is an aggregate of VCIs.
- *Payload type* (PT)—the 3-bit PT field indicates whether the cell payload contains user or network management information.
- *Cell loss priority* (CLP)—the 1-bit CLP field indicates whether a cell can be discarded if the network becomes congested. If the CLP is set to one, the cell is subject to discard.
- *Header error control* (HEC)—the HEC uses an 8-bit error code to correct single-bit errors in the header and to detect double-bit errors.

The information field (48 bytes) carries the payload within the cell.

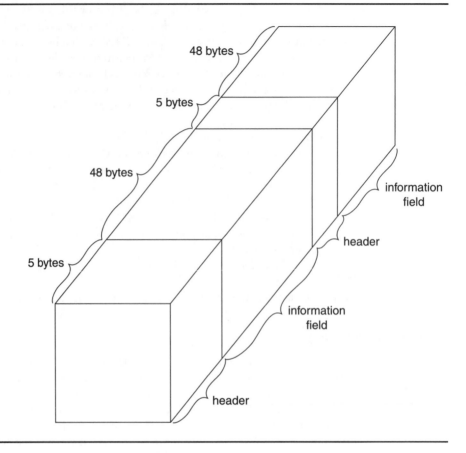

FIGURE 8.5   Stream of ATM cells.

ATM Standard Layers

The ATM standard defines three layers: the physical layer, ATM layer, and ATM adaptation layer (Figure 8.6).

Physical Layer

ATM cells may be transported over many different physical media and still maintain compatibility. The physical layer defines how ATM cell streams are transmitted over the physical media, as well as the interface to the ATM layer. Cells are transported within the ATM layer either asynchronously, as in packet–switching, or synchronously, as payloads encapsulated in SONET envelopes (Figure 8.7).

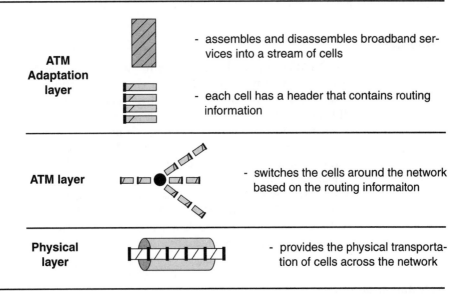

FIGURE 8.6   ATM layers.

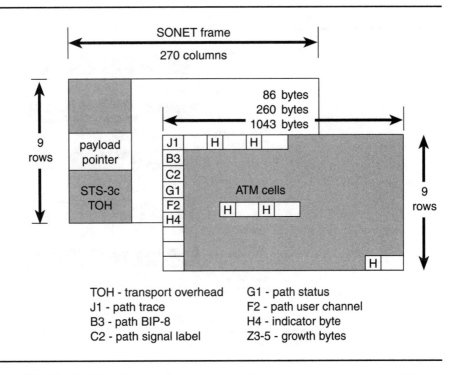

TOH - transport overhead   G1 - path status
J1 - path trace   F2 - path user channel
B3 - path BIP-8   H4 - indicator byte
C2 - path signal label   Z3-5 - growth bytes

FIGURE 8.7   ATM Physical layer transport.

ATM Layer

The ATM layer provides the switching capability for the ATM cells by means of virtual connections. Two kinds of virtual connections have been standardized: virtual channel connections and virtual path connections. By reading the VPI and VCI bits in the header of each cell, an ATM switch routes cells to their destination. Virtual channels are grouped together to form a virtual path. Many virtual channels may share a single physical link at the same time. For example, all the virtual channels that belong to a customer may be bundled within a single virtual path, simplifying network management. A virtual channel may have other attributes, such as quality of service, associated with it. Should congestion occur, the ATM switch selectively drops cells until the congestion clears. (Overload conditions may also be handled by a policing mechanism which limits services to their negotiated bandwidth.) The selection of which cells to lose is based on the guaranteed quality of service.

ATM Adaptation Layer

Above the cell-switching layer are *ATM adaptation layers* (AAL), which map various kinds of traffic into and out of the cells (Figure 8.8). The adaptation

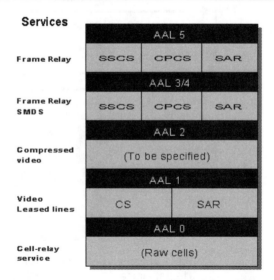

CPCS = Common-part convergence sublayer   SMDS = Switched multimegabit data service
CS = Convergence sublayer   SSCS = Service-specific convergence sublayer
SAR = Segmentation and reassembly

FIGURE 8.8   ATM adaptation layers.

TABLE 8.1   ATM adaptation layer.

Feature	AAL1	AAL2	AAL3	AAL4	AAL5
Timing relation between source and destination	required	required	not required	not required	not required
Bit rate	constant	variable	variable	variable	variable
Connection mode	connect-oriented	connect-oriented	connect-oriented	connection-less	connect-oriented

layers must differentiate between data, voice, and video traffic because of their very different transmission requirements. There are five types of AAL as defined in the ATM Standard that are specialized to types of traffic (Table 8.1).

1. AAL1 is a constant bit-rate service for voice and video traffic.
2. AAL2 is a variable bit-rate service for audio and video.
3. AAL3 is a connection-oriented service for data.
4. AAL4 is a connectionless service for data.
5. AAL5 is a high-performance multimedia service.

The AAL continues to evolve. The earlier AAL3/4 supports native BISDN, while the later, AAL5 was constructed to support the use of BISDN for the transport of existing protocol services at a higher performance level. AAL5 services include:

- Notification of corrupted received AAL *protocol data units* (PDUs)
- Unverified data transfer (error recovery in higher layers)
- PDU transfer from one AAL to another

AAL5 could bring ATM to the desktop for distributed database services, computer-aided design, and manufacturing, and could facilitate high-speed multimedia applications.

To provide a particular service, ATM maps the service into the information field of a cell. When the cell is full, the correct VCI/VPI information is placed in the header field and the cell enters the ATM cell stream. The AALs have two logical sublayers known as the *convergence sublayer* (CS) and the *segmentation and reassembly sublayer* (SAR) that support this process. The CS accepts data units from the ATM user interface and delivers the data units back after receiving them from the SAR. The SAR divides each data unit into cells on the segmentation side and reconstructs incoming cells into data units on the reassembly side.

The CS ensures that the different types of traffic receive the right level of

user data (up to 64K bytes)	PAD (0–47 bytes)	control field (2 bytes)	length field (2 bytes)	CRC-32 (4 bytes)

FIGURE 8.9   Protocol data unit structure.

service at the *user-to-network interface* (UNI) and at the *network-to-network interface* (NNI). The CS passes PDUs on to the network from the UNI and delivers PDUs back after reception (Figure 8.9). The SAR sublayer divides each PDU into cells during segmentation and reconstructs incoming cells into PDU during reassembly. When transmitting into the network, the CS appends a PDU trailer and pads the PDU plus trailer in multiples of 48 bytes. The trailer contains a cyclic redundancy check field (CRC-32), a PDU length field, and a control field. Cells formed during the AAL layer segmentation process will all be 48 bytes. The SAR reassembles incoming cells into complete PDUs before passing them to the CS. Information for the reassembly is obtained from a combination of the VPI/VCI fields and the end-of-message indicator encoded in the PT field. After the CRC-32 is verified, the PDU is delivered to the ATM user.

STANDARD AND FORUM GROUPS

The development of any communications standard demands global coordination. Broadband network standards are created by a diverse group of standardization bodies and industry forums, all involved with the task of generating a consensus view on implementing a communications technology (Figure 8.10). The groups are continually refining, extending, and adding more advanced capabilities. For example, the X.25 Packet Network Standard, initially released in 1974, continues to be updated every four years. At each revisit the cumulative experience with the older version is used to update and improve it.

International Telecommunication Union (ITU)

Formerly the CCITT, the ITU provides communications standards for the United Nations. Its study groups have specific technical responsibility for developing international standards (for example, ATM and SONET/SDH standards

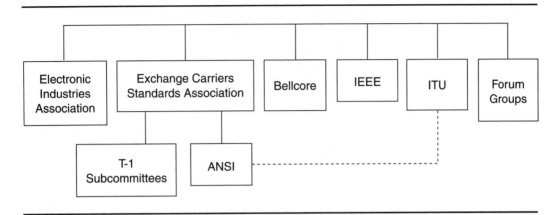

FIGURE 8.10   Broadband standard—setting organizations.

are being developed in Study Group XVIII). Every four years the study groups submit recommendations to the ITU plenary for approval. While the standards that are approved by the ITU are law in some European Countries, in most nations they are treated as recommendations. The ITU's influence in setting standards comes from its coordination with each national standards group (e.g., ANSI), the thoroughness of its standards process, and its worldwide stature.

American National Standards Institute (ANSI)

ANSI was created in 1918 for the purpose of coordinating private-sector standards development in the United States. It is also the U.S. representative to international standards groups and includes hundreds of standards committees as well as associated groups such as the *Exchange Carriers Standards Association* (ECSA). The ANSI subcommittees that are active in broadband network standards are T1X1 and T1M1. T1X1 plays a role in SONET rates and format specification, while T1M1 guides the effort to define the standard for *operations, administration, maintenance, and provisioning* (OAM&P).

Forums

A recent phenomenon among standards groups is the emergence of forums from various commercial groups. The purpose of a forum is to impact the standards process in a positive way and on the behalf of a particular communications technology. A forum consists of a group of interested carriers and vendors who support relevant education and implementation activities.

ATM Forum

The ATM forum is an international consortium of members that is chartered to accelerate the acceptance of ATM products and services in local, metropolitan, and wide-area networks. Formed in October 1991, the ATM Forum membership currently includes more than 700 companies representing all sectors of the computer and communications industries, as well as a number of government agencies, research organizations, and users. Although it is not an official standards body, the ATM Forum works with the official standards groups, ANSI and the ITU, to ensure interoperability among ATM systems. The consortium accelerates ATM adoption through the development of common implementation specifications. The Forum's first specification on the ATM UNI provides an important platform upon which vendors can design and build equipment. It defines the interface between a router or a workstation and a private ATM switch, or between a private ATM switch and a public ATM switch. The ATM Forum has created interest in the private use of ATM in LANs. Some of their activities include promoting LAN management standards such as SNMP for ATM network management and encouraging the use of small switched ATM networks as LANs.

Frame-Relay Forum

The Frame-Relay Forum, an organization open to vendors and users, is dedicated to promoting the acceptance, implementation, and interoperability of frame relay based on national and international standards. It was incorporated in May 1991 as a nonprofit mutual corporation and has over 300 members worldwide. Offices are located in North America, Australia/New Zealand, Europe, and Japan.

Other Groups: IEEE and IETF

There are many other groups associated with broadband network standards. *The Institute of Electrical and Electronic Engineers* (IEEE) is very active in LAN standards. The *Internet Engineering Task Force* (IETF) is concerned with the TCP/IP protocol suite and has released standards that enable data communications between TCP/IP and ATM networks.

CONCLUSION

With hundreds of millions of desktop computers connected by worldwide LANs, it was natural for the LANs themselves to be interconnected, first in geographically close areas, later in metropolises, finally in nations and around the globe. Not just LAN computer traffic was infused into the WAN, but computer technology as well. The LAN topologies, high bandwidths, stripped-down protocols, open architectures, and management were adopted by the WAN. New broadband technologies such as frame and cell relay, SONET/ SDH, and ATM have left the standard bodies and entered the Telco/PTT networks. ATM, for example, allows the Telco/PTT to tailor a variety of services to the specific needs of business and residential customers. Now the same technologies, honed in the WAN, are returning to the LAN mainstream to route information at unprecedented speeds.

Broadband Transport Technologies

INTRODUCTION

The widespread use of digital technology in the WAN occurred because bandwidth was—and still is—expensive. Digital technology provides the same economies in communications as it does in computers, making better use of available bandwidth and allowing more complex services and equipment to be offered at lower cost. Encouraged by favorable tariffs, users have upgraded to digital services. This growth is accelerating because of the benefits offered by advanced digital data communications protocols. These protocols, first deployed in the computing environment, are being modified to create synchronous overlays to the global asynchronous telephone network.

The Telco/PTT use of asynchronous protocols derives from worldwide networking standards implemented over the past 40 years. These protocols operate at the physical level (OSI layer 1) and are transparent to the computer communication protocols they transport. In North America, the standard for high-speed WAN interconnection is the T1, with a fixed bit rate of 1.544 Mb/s. This rate supports 24 digital channels of 64 kb/s each, plus 8 kb/s for signaling. At the time of the introduction of T1 service, 64 kb/s was the bandwidth needed to transport a single voice conversation, using a method of modulation called *pulse code modulation* (PCM). Today, toll-quality voice can be compressed even more, allowing a 64 kb/s channel to carry multiple voice conversations. Nonetheless, the fundamental rate for voice transmission and switching in the public telephone network remains 64 kb/s.

Within the public telephone network there is a scale of rates that goes beyond T1, referred to as the *asynchronous standard hierarchy*. While consistent at each level, it has little overlap, so asynchronous DS1 frames have little resemblance to DS3 frames. (T1 equals 1 DS1; T2 equals 4 DS1; T3 equals 28 DS1.) As a result, the public network is synchronous only on a piecemeal basis, and therefore lacks the management capability and bandwidth flexibility for many newly demanded services. In contrast, LANs provide inexpensive connectivity, at least in a local area. But LANs, too, have come under pressure from the increasing bandwidth demands of new applications.

The result is a tenfold increase in bandwidth and use of fiber media. FDDI, a dual counter-rotating token ring topology LAN operating over fiber, is being promoted as a solution for broadband applications. But shared-media LANs of any variety including FDDI are intrinsically limited. Video, as well as voice traffic, tolerates relatively small delays and requires synchronization signals that shared-media LANs lack. In campus networks, for example, video cannot be effectively transported, even with compression as low as 128 kb/s. If there were only a few workstations on an FDDI LAN, such compression might be enough, but that is not the case in a shared-media environment.

BROADBAND ASYNCHRONOUS TRANSMISSION: T3/E3

T3, an asynchronous transmission service, represents the equivalent of 28 T1 lines at a rate of 44.736 Mb/s. Although offered by Telcos/PTTs, there are issues to consider before deploying T3 in corporate networks. First, asynchronous signal types bear no relationship to each other. Within a 45 Mb/s DS3 signal, the 1.5 Mb/s DS1 signal has no visibility. The entire DS3 has to be taken apart to reach the DS1. Then the signal has to be reframed, taking time and processing power. Second, DS3 is often transmitted over fiber, which requires an interface for electrical-to-optical signal conversion. Unfortunately, asynchronous transmission only provides a standard electrical interface. The lack of an optical standard for asynchronous signals has led to many proprietary interfaces. (T3 services require that the customer select with the carrier the type of optical interfaces to be placed in the interexchange carrier's serving office.) These proprietary optical transmission rates include 45 Mb/s (typically 48 to 50 Mb/s), 90 Mb/s, 135 Mb/s, 405 Mb/s, 560 Mb/s, 565 Mb/s, and 1.1 Gb/s— all of which are unique to individual terminal equipment vendors. The T3 rate of 44.736 Mb/s represents the North American standard. The Japanese T3 equivalent is 32 Mb/s, while the European equivalent is E3, which operates at 34 Mb/s and supports E1 signals at 2.048 Mb/s transmission rates. Both E3 and T3 are insufficient for intelligent networks. Although there is some capability for maintenance signaling, T3/E3 actually have less maintenance visibility than lower-rate T1/E1.

The situation is improving. Rather than jury-rigging the asynchronous standards for improved management capability, a more general approach has been taken by the worldwide standards bodies. Networks have begun the migration from asynchronous to synchronous transmission. With the almost inexhaustible bandwidth availability of fiber cable, the silicon integrated circuit revolution, and the improved synchrony of network clocks, both packet-switched and isochronous traffic can be transported synchronously over large distances. Today, a variety of networking technologies is available to address the rapidly emerging class of bandwidth-ravenous applications. These broadband networking technologies include the fast-packet schemes, such as frame and cell relay (ATM) previously discussed, and SONET/SDH. SONET/SDH represents a transmission format that includes in-band network management. It overlays the existing asynchronous broadband network and will eventually replace it.

SONET AND SDH

The asynchronous standards specify the termination and aspects of a limited amount of network management. Access to the network is specified as a standard electrical interface for connection compatibility. For DS3, the rate of 44.736 Mb/s is defined, together with a signal level, impedance, and other characteristics. This is not adequate to achieve interoperability among equipment vendors (Figure 9.1). Both SONET and SDH standards specify the electrical and optical interfaces as well as protocols at different points on the communication line, permitting the use of equipment from different vendors anywhere along the fiber span (Figure 9.2). This capability is referred to as *mid-span meet*. Customers get better bandwidth control as well as reduced Telco/PTT costs because SONET/SDH–com-

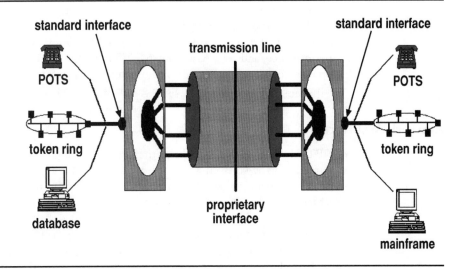

FIGURE 9.1   Asynchronous connection.

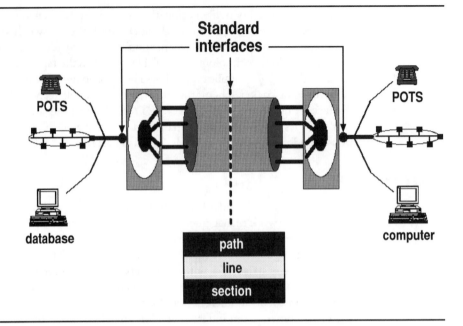

FIGURE 9.2   Synchronous connection by SONET.

pliant equipment is interoperable. Furthermore, its ability to concatenate signals consistently at any speed makes multiplexing sufficiently low-cost to be embedded in fiber-optic equipment. Given this, real-time switchable multiplexers, add/drop systems, and ring networks can now be built for improved network reliability, performance, and cost. As a result, SONET (in North America) and SDH (in Europe) deployment is gaining great momentum and could reach parity with the asynchronous market by 1997, equating to many billions of dollars.

Network Architecture

SONET/SDH technology combines the historical elements of LANs and WANs with much borrowed from advanced computer technology. The SONET in-band network management ASN.1 protocol derives from the OSI standards on object language definition that was developed for computers. This characterization by a communications standards body reflects the coalescing of LAN and WAN technology. Due to its powerful architecture, SONET integrates existing point-to-point fiber links into true networks, enabling them to route single-voice grade channels digital service level Ø (DS0) without the need for multiple stages of multiplexing and demultiplexing.

The eventual role of SONET and fiber systems will not be simply to replace the dedicated copper lines used in transmitting multiplexed digital signals. SONET networks will simultaneously offer greater functionality, standardized fiber interfaces, higher transmission speeds, and reduced maintenance. SONET provides a flexible, controllable network with centralized network management. Because of its capacity to manage large amounts of bandwidth, as well as its simplicity and cost-effectiveness, it will eventually halt purchases of nonstandard fiber and asynchronous communications equipment. It already is the preferred delivery vehicle for a number of emerging broadband services, including frame relay, SMDS, BISDN, and ATM (Figure 9.3).

For this vision of broadband networking to be fully realized, the Telcos/PTTs will have to upgrade their infrastructure to support SONET signals that operate at higher data rates than their asynchronous counterparts. Although the asynchronous network is unable to support SONET signals, SONET will transport asynchronous signals such as DS1, DS1C, DS2, and DS3. Therefore, existing asynchronous equipment such as multiplexers, DACs, and switches will not have to be immediately replaced as SONET equipment migrates into the public network.

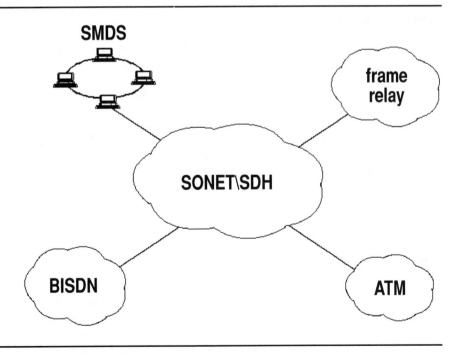

FIGURE 9.3   SONET transport for emerging services.

TABLE 9.1 SONET rates and bandwidths.

OC Optical	STS Electrical	Rates (Mb/s)		
		Line	Payload	Overhead
OC–1	STS–1	51.840	50.122	1.728
OC–3	STS–3	155.520	150.336	5.184
OC–9	STS–9	466.560	451.008	15.552
OC–12	STS–12	622.080	601.344	20.736
OC–18	STS–18	933.120	902.016	31.104
OC–24	STS–24	1244.160	1202.688	41.472
OC–36	STS–36	1866.240	1804.032	62.208
OC–48	STS–48	2488.320	2405.376	82.944
OC–96	STS–96	4976.640	4810.752	165.888
OC–192	STS–192	9953.280	9621.504	331.776

Transmission Rates

Using a basic building block called the *synchronous transport signal level-1* (STS-1), with a line rate of 51.840 Mb/s for SONET and 155.52 Mb/s for *synchronous transport mode level-1* (STM-1) for SDH, SONET/SDH reaches 2.488 gigabits per second in exact multiples of STS-1. Table 9.1 shows the relationship of line rates, payload, overhead bandwidths, and the corresponding STS levels. (SONET, a North American ANSI and Japan standard, uses the STS signal hierarchy. The European ITU equivalent, SDH, uses the STM signal hierarchy.) STS electrical signals, when transmitted over fiber, are converted to a corresponding optical signal called *optical carrier* (OC). For example, higher SONET transmission rates are established by concatenating "N" STS-1s to form an STS-N. Currently, "N" is defined as 1, 3, 12, 24, 36, and 48, but rates up to OC-255 or 13.2192 Gb/s are possible.

Synchronous Clocks

Unlike asynchronous signals, the synchronous signal is not demultiplexed and then remultiplexed at every central office through which it passes. SONET allows simple add/drop multiplexing. That is, information as fine as a single DS0 can be taken from one broadband data stream and inserted into another. This process requires clock synchrony within the public telephone network. When the asynchronous public network was designed, such clocks did not

TABLE 9.2   Clock Hierarchy for the Public Telephone Network

Stratum	Accuracy	Slip Interval
1	.01 part in one billion	72 days
2	16 parts in one billion	14.5 days
3	46 parts in one billion	5.6 min.
4	32 parts in one million	3.9 sec.

exist. Delays due to the vacuum tubes and transistors used then were too great. SONET network clocks may be distributed across thousands of kilometers, making synchronous transmission difficult to maintain. The issue occurs because of the way that the clocks are formed.

The clocks in the synchronization network are classified on the basis of performance. There are four levels, called strata (Table 9.2). *Stratum 1* (ST1) is the highest level and its free running clocks are used as *primary reference clocks* (PRC), compared with ST3 clocks that are normally locked to an incoming reference, which is traceable to an ST1.

To optimize the data density, digital fiber-optic systems carry data in *non-return to zero* (NRZ) format: A light pulse corresponds to a logical "1"; and the absence of a pulse is a logical "0." This is very efficient because clock frequency and the NRZ data transfer rate are the same. But, synchronization is difficult at high data rates because the clock is implicit and the receiver needs to extract the clock before any signal processing can be done. However, the data signal's frequency spectrum contains no component at the carrier frequency. Moreover, although the transmit clock has a known nominal frequency, it will wander due to unequal tolerances, temperature variations, and aging. A typical frequency accuracy specification for telecommunications fiber-optic systems is 20 PPM plus or minus. As a result, the receiver cannot exactly predict the clock frequency.

Originally, SONET was considered immune to timing anomalies because of its use of pointers. The pointer process was designed to accommodate synchronization variations. But it has, instead, created a new type of timing anomaly. The public telephone network BITS clock systems were designed around standards developed to support asynchronous networks as well as networks that receive their timing from several PRC sources (plesiochronous networks). Originally, areas such as short-term clock stability and phase transients did not affect network operation. But pointers can increase to an intolerable level with the slightest timing variation. When this occurs, an 8-bit or unit interval phase discontinuity is transmitted by the SONET terminal multiplexer to the asynchronous tributary. These phase steps could create severely erred seconds on DS1 traffic. Thus, the vendor requirements of Bellcore and ANSI specify extremely stringent limits on the stability of a timing feed. This is satisfied by a PRC, but may not be obtainable from the existing network clocking scheme (see Chapter 10).

LAYERED STRUCTURE

The overhead is divided into three layers—path, line, and section—that govern the transport of the payload across the network (Figure 9.4). The relationship between SONET layers plays a very important role with regard to the various network services, DS1, DS3, and so forth, that it supports.

Physical Layer

The physical layer transports bits as optical pulses through the fiber medium. No overhead is associated with this layer. The main function of the physical layer is to receive the STS-N signal bit stream, convert each bit into an optical pulse, and transmit the optical pulse across the fiber medium towards the far-end terminal. Key concerns are pulse shape, power levels, and the line code for error detection and recovery.

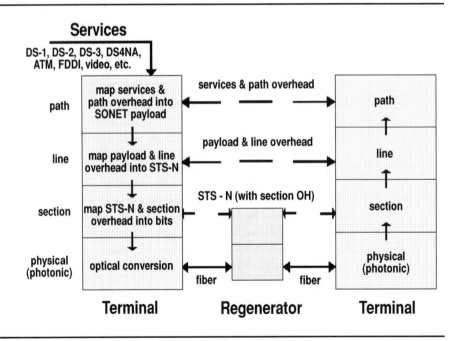

FIGURE 9.4   Layered networks.

Section Layer

The section layer transports the STS-N frame across the physical medium. It uses the physical layer for transport. Section layer functions include framing, scrambling, section overhead processing, and error monitoring. The overhead defined for this layer is interpreted or created by *section terminating equipment* (STE).

Line Layer

The line layer is responsible for transporting the payload and the line overhead to its peer at the far end. All lower layers provide transport. This layer maps the payload and the line overhead into STS-N frames. The payloads and line overheads are synchronized and multiplexed within the STS-N, before the STS-N signal is passed to the section layer. The overhead associated with this layer is interpreted or created by *line terminating equipment* (LTE).

Path Layer

The transport of network services between two SONET multiplexing nodes is the function of the path layer. Examples of such services are the provisioning of DS1, DS2, DS3, DS-4NA, FDDI, ATM, video, and so forth. The path layer maps the services into the SONET payload format as required by the line layer and communicates end-to-end via the path overhead. The *path terminating equipment* (PTE) interprets or creates the overhead defined for this layer.

SECTION, LINE, PATH STRUCTURE

Section, line, and path are structures that delineate portions of the fiber-optic transmission facility. These interconnect the SONET network elements (Figure 9.5).

Section

Section is the segment of the SONET transmission facility that includes terminating points between a network element and one or two regenerators.

Line

Line, with LTE, transports information between two consecutive line-terminating network elements; one that originates and another that terminates the line signal. In Figure 9.5 this represents the sections between the *add/drop multiplexer* (ADM) or *digital cross-connect system* (DCS) and the *terminal multiplexer* (TM).

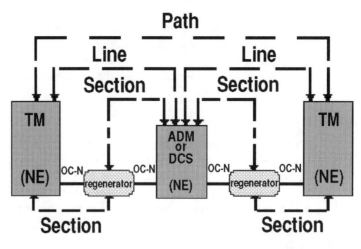

TM = terminal multiplexer
DCS = digital cross-connect

ADM = add/drop mutiplexer
NE = network element

FIGURE 9.5   Section, line, and path.

Path

A path is defined as a logical connection between two points: one (source) at which the frame is assembled and another (sink) where it is disassembled, as in the link between the two TMs in Figure 9.5.

SONET FRAMES

The lowest data rate supported by SONET/SDH is the 51.840 Mb/s STS-1 frame. An STS-1 frame is a 9-row-by-90-column byte matrix structure totaling 810 bytes or 6,480 bits (Figure 9.6). The byte from row 1, column 1 is transmitted first, then is followed by row 1, column 2, and so on. The transmission is from the left column to the right column and from the top row to the bottom row. The STS-1 frame transmission duration is 125 microseconds, or 8,000 frames per second, which maintains compatibility with the existing telephone network. The first three columns, or 27 bytes, are assigned to the transport overhead, which is subdivided into section overhead (9 bytes) and line overhead (18 bytes). The remaining 9 rows by 87 columns constitute the STS-1 *synchronous payload envelope* (SPE) that has a total of 783 bytes. Of these, 9 bytes

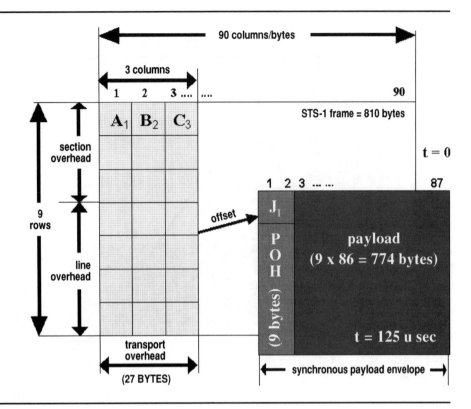

FIGURE 9.6   STS-1 format.

in the first column are designated as STS path overhead or POH. The actual payload is 774 bytes, which results in a total of 49.536 Mb/s of payload capacity.

This payload capacity is used to carry a DS3 signal at 44.736 Mb/s, or 28 DS1s each, at 1.544 Mb/s. All this is at the STS-1 level. For the transport of payloads with bandwidth requirements greater than STS-1, several STS-1s can be combined or concatenated (byte interleaved) and transported as a single entity. For example, the ITU has defined the SDH lowest-rate STM-1 as the equivalent to SONET STS-3.

Before byte interleaving, the STS-1s are frame-aligned so the transport overhead of each STS-1 can be combined to form an STS-N transport overhead. Thus, all transport overheads are frame-aligned, while the individual payloads float within the envelope as indicated by the respective payload pointers within each STS-1 transport overhead. The byte-interleaved STS-N signal, when transmitted, results in an optical signal called Optical Carrier OC-N, with

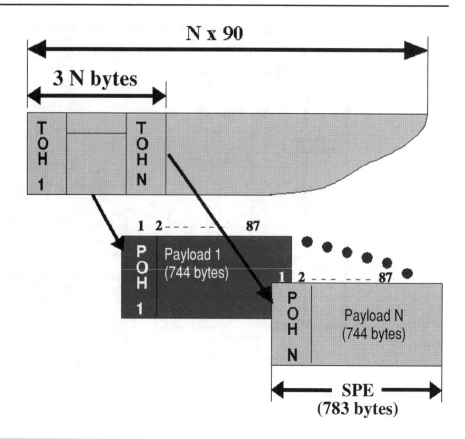

FIGURE 9.7   STS-N frame format.

values of N. The overhead byte C1 in the section overhead identifies the STS-1 within an STS-N frame format.

The *transport overhead* (TOH), together with section and line overheads, performs the functions needed to transport the SPE over the fiber link. Also, the payload pointer or *offset* resides within the line overhead of the TOH. The payload pointer plays a very crucial role in the transmission process because it indicates the start of the STS-1 SPE. It also permits the payload to float within the envelope and helps adjust frequency deviations between network elements. A 9-byte *path overhead* (POH) is allocated within the synchronous payload envelope to support transport of the payload from the point at which it is assembled to the point at which it is disassembled.

The size of an STS-N frame is N times the STS-1 frame size or N x 810 bytes (Figure 9.7). Similarly,

Transport overhead = $N \times 27$ bytes

Synchronous payload envelope = $N \times 783$ bytes

Path overhead = $N \times 9$ bytes

Payload size = $N \times 774$ bytes

STS-Nc Concatenation Format

The structure for STS-Nc concatenation is illustrated in Figure 9.8. SONET accommodates higher transmission rates by synchronously byte-interleaving 'N' STS-1s to form an STS-N signal. Each STS-1 within the STS-N, however, remains a separate entity that is assembled at the source and disassembled at the sink point. The size of an STS-Nc is $N \times 810$ bytes, while the transport overhead is $N \times 27$ bytes. Both overhead and payload are treated as single entities. Only one POH is shown. The remainder is allocated to the payload that, in this case, is ($N \times 783 - 9$) bytes. Moreover, in the STS-Nc frame format, the first STS-1 carries a normal pointer in its transport overhead, while the pointers of other STS-1s, forming the STS-Nc, carry the concatenation indicator. This indicator helps in binding the constituent STS-1s together and multiplexing, switching, and trans-

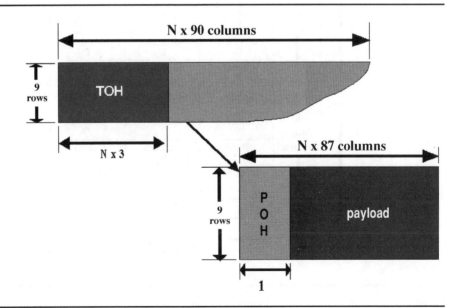

FIGURE 9.8   STS-Nc concatenation format.

porting them as a single entity. A concatenation indicator in the transport overhead shows that the STS-1s of an STS-Nc are joined—and will remain—together until terminated. By byte-interleaving 'N' STS-1s, SONET provides for super rate services, such as BISDN which uses multiples of the STS-1 rate.

Section, Line, and Path Overheads

In the STS-1 frame format, the TOH consists of 27 bytes of which 9 are assigned to section overhead and 18 to line overhead. These bytes occupy the first 3 columns of the STS-1 frame. The POH has 9 bytes that form the first column of the SPE. The section overhead is created or modified within the section segment, the line overhead within the line segment, and the path overhead within the path segment. Each layer, section, line, or path processes the associated overhead prior to being inserted in the signal. Each overhead assists the transport of the STS-N signal between two end points of the transmission segment. Figure 9.9 shows the details of section, line, and path overheads.

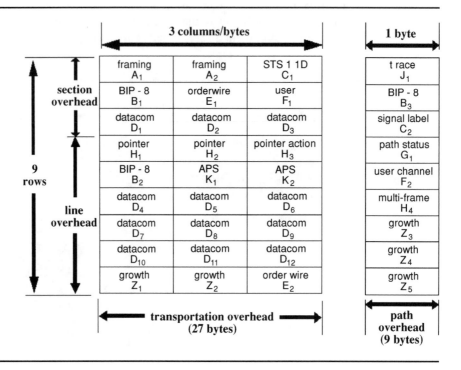

FIGURE 9.9 Section, line, and path overheads.

Section Overhead

- A1 and A2 bytes provide STS-1 framing synchronization.
- C1 byte identifies each STS-1 within an STS-N.
- B1 byte is used for section error monitoring.
- E1 provides a local channel for voice communications.
- F1 byte is reserved for the network provider.
- D1, D2, and D3 bytes provide a 192 kb/s *data communications channel* (DCC) for section operation, administration, maintenance, and provisioning.

Line Overhead

- H1 and H2 bytes are used as pointers to the start of the STS-1 synchronous payload envelope (SPE).
- H3 byte provides for SPE frequency adjustment in conjunction with H1 and H2 bytes.
- B2 is used for line error monitoring.
- K1 and K2 bytes provide automatic protection switching.
- D4 to D12 bytes form the 576 kb/s data communications channel for line operations, administration, maintenance, and provisioning.
- Z1 and Z2 bytes are reserved for future growth.
- E2 provides an express orderwire channel between line entities.

Path Overhead

- J1 byte is used to verify connection continuity between the receiver/transmitter pair.
- B3 is used for path error monitoring.
- C2 indicates the equipped/unequipped status of STS-1.
- G1 byte is used to monitor status and performance of the end-to-end path.
- F2 is reserved for the network provider.
- H4 byte indicates phases within the STS-1 frame.
- Z3, Z4, and Z5 bytes are reserved for future growth.

This overhead structure provides powerful fault-detection, sectionalization, and reconfiguration capabilities. The SONET/SDH path and line section manage heterogeneous environments, providing end-to-end control by means of the frame overhead bytes. The actual overhead byte values depend upon how the equipment will be used. For instance, the K1 and K2 bytes transport messages between the network elements and the network management system. With a ring topology, a more sophisticated K1/K2 protocol would be required to switch traffic direction, since the traditional alarming protocol of the telephone network would be incompatible with a ring architecture. (This topology was previously used in LANs rather than WANs because of the expense of redundant paths and because the alarm protocols of the PSTN such as *far-end receive* had to be reworked.)

The *embedded operations channel* (EOC) within the DCC allows remote control of the operations, administration, maintenance, and provisioning functions for interconnected SONET network elements, thereby reducing the need to dispatch technicians. The EOC protocol follows the seven-layer model for OSI and thus permits internetworking with intelligent OSI network controllers.

Virtual Tributaries

SONET framing accommodates lower-rate signals, providing transport for existing North American and international formats. The STS-1 payload may be subdivided into smaller *virtual tributaries* (VTs) that transport signals at less than DS3. Each VT functions as a separate container within the STS-1 signal with its own overhead bits. Because SONET/SDH is an international standard, the most common North American (DS1 at 1.544 Mb/s) and international (E1 at 2.048 Mb/s) tributaries have defined VT mappings. Less common tributaries, such as DS1C and DS2, are also represented. There is a special mapping for higher-rate DS3 signals within a SONET payload—and eventually 10 Mb/s for Ethernet and 16 Mb/s for token ring LANs.

There are both *locked* and *floating*, *channelized* and *unchannelized* VT modes. The unchannelized floating mode is for wideband cross-connection down to the DS1 level. In unchannelized operation, the lowest addressable level is a VT. Two unchannelized mappings have been defined: asynchronous and bit synchronous. The former requires minimum timing consistency between the tributaries and the SONET clock, whereas the latter establishes a common clock frequency, but assigns an arbitrary phase. The locked mode fixes the VT location within the SPE that supports DS1 cross-connects. DS0 channels are uniquely addressable within a SONET payload by means of channelization. This is also known as *byte-synchronous* operation, in which both a fixed clock frequency and a fixed phase are established based upon the DS1 frame. This enables individual DS0s (8-bit bytes) to be easily identified and cross-connected. The channelized locked mode provides for DS0 level cross-connection and bridging into an existing DS3 network. The floating mode employs pointers that define the location of the VT. An advantage of VT pointers is that they remove the need for slip buffers. In SONET equipment, the synchronizer, desynchronizer, and slip buffer are replaced by a pointer processor.

A VT group architecture accommodates the mixes of various VT types within an SPE. The VT group size is a constant 9 rows by 12 columns or 108 bytes. Mixing VT types within the same group is not allowed, but the number of VTs within a VT group depends on the VT size. For example, four VT-1.5s, three VT-2s, two VT-3s, or one VT-6 may be packaged into a single VT group and seven VT groups may be byte-interleaved with the overhead to form the SPE.

SONET Standards

Lightwave fiber-optic digital transport systems have been used for decades. Non-standard or proprietary transmission rates of 90, 140, 405, 565, and 810 Mb/s, and 1.2 and 1.7 Gb/s, are deployed within these optical communications networks. The ECSA T1 committee was established in 1984 to coordinate the development of U.S. telecommunications standards. The ANSI fully supports this ECSA T1 committee. The SONET specifications comply as well with the ITU, formerly the *Consultative Committee on International Telephony and Telegraphy's* (CCITT), SDH recommendations. The ITU is the United Nations body that makes recommendations and coordinates the development of telecommunications standards for the entire world. Although there are some differences between the two standards, particularly in the areas of rates and management, they are mostly compatible.

SONET, the North American standard, and SDH the international standard, are being cooperatively developed as a result of their complexity and worldwide scope. The standards are being released in phases. This allows vendors to design equipment that can be readily upgraded to conform to the emerging SONET/SDH standards. The Phase I Standard was released in 1988 by the T1 committee.

Phase I

Phase I has two components: T1.105 and T1.106. T1.105 defines as:

- Byte-interleaved multiplexing format
- Line rates for STS—1, 3, 9, 12, 18, 24, 36, and 48
- Mappings for DS0, DS1, DS2, DS3
- Monitoring mechanisms for section, line, and path structures
- 192 kb/s and 576 kb/s DCC

T1.106 also specifies the optical parameters for the long-reach, single-mode fiber cable systems.

Phase II

Phase II standards, released during 1990/1991, have three components, namely T1.105R1, T1.117, and T1.102-199X. T1.105R1 is a revision of the earlier T1.105 and specifies:

- SONET format clarification and enhancements
- Timing and synchronization enhancements
- *Automatic protection switching* (APS)
- Seven-layer protocol stack for DCC and embedded operations channels
- Mapping of DS4 (139 Mb/s) signal into STS-3c

In addition, T1.117 specifies the optical parameters for short-haul (less than two kilometers) multimode, fiber cable systems, and T1.102-199X gives electrical specifications for STS-1 and STS-3 signals.

The Phase I and II specifications have sufficient detail to initiate development of SONET-integrated circuit chip sets by device vendors. The vendors are using several silicon technologies to cope with the wide spread in bit rates. At the low end (up to 622 Mb/s or OC-12), CMOS, BiCMOS, and ECL gate arrays satisfy the performance requirements. Gigabit-per-second speeds require high-powered, submicron bipolar, and Gallium-Arsenide (GaAs) technology, which is more costly. Because of integrated circuits, network elements such as ADMs, DCSs, and *remote digital terminals* (RDT), among others, are commercially available.

Phase III

Phase III addresses synchronization, ring topologies, network survivability, service mapping, operations, administration, maintenance, and provisioning message sets, and DCC/LAN standardization and addressing (Figure 9.10).

Phase III consists of four components, T1.105.01, T1.105.05, T1.105.03, and T1.119.

- ANSI T1.105.01-1994 provides the requirements for two fiber and four fiber bi-directional line-switched SONET rings. These rings allow failure restoration within 50 milliseconds and are an important part of many network restoration plans.
- ANSI T1.105.05-1994 describes the Tandem Connection Overhead layer for SONET. It allows carriers to maintain groups of STS-1 signals as a single unit and offers users improved monitoring and service.
- ANSI T1.105.03-1994 provides the jitter requirements for SONET interfaces and at interfaces between SONET and asynchronous networks. These SONET jitter specifications help ensure adequate performance at SONET network interfaces.
- ANSI T1.119-1994 allows the management of SONET network elements using an OSI compliant interface. This will be followed by a standard for SONET operations, administration, maintenance, and provisioning.

There are still issues to be resolved, including:

- Expanding the SONET Standard to include the OC-192 rate of 13 Gb/s
- Defining interfaces at rates lower than 50 Mb/s
- Developing an OSI-compliant directory services standard based on X.500 that will allow for the automatic registration of SONET network elements
- Enhancing SONET rings to include low-priority traffic access to protec-

Completed Standards

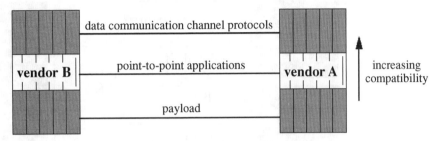

data communication channel protocols

vendor B　　point-to-point applications　　vendor A

payload

increasing
compatibility

Uncompleted Standards

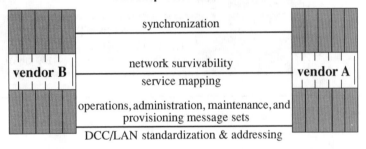

synchronization

vendor B　　network survivability

service mapping

vendor A

operations, administration, maintenance, and
provisioning message sets

DCC/LAN standardization & addressing

FIGURE 9.10　　SONET standardization.

tion channel, virtual tributary ring access, and multinode ring interconnection for improved reliability
- FDDI mapping so that a 125 Mb/s FDDI signal maps into a 155 Mb/s STS-3c SONET signal

CONCLUSION

The SONET embodiment was conceived at Bellcore in 1984. SONET standards remain in development and are arriving in three phases. Phase I standards, issued in 1988, have permitted the development and testing of early SONET equipment. Phase II standards were issued in 1991. These standards had enough detail for vendors to develop commercial SONET equipment. Despite being extremely complex, portions of the Phase III definition have been completed. Chip and device designers have produced standard components conforming as

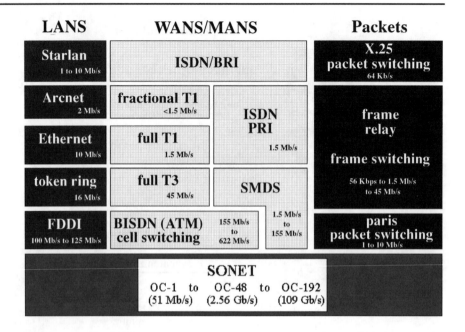

FIGURE 9.11 SONET service infrastructure.

closely as possible to SONET requirements. Due to the changes in the standards, most vendors have taken into account the need for future reengineering to meet any unresolved fundamental specifications. Higher-level product developers are using both standard and proprietary devices to build multiplexers, digital cross-connects, and remote terminals. All have their own individual sets of plans to field retrofit for compliance with still undefined standards. Still, some vendors are having difficulty with properly supporting their products which is why customers must actively continue to exercise "vendor management" and be aware of the SONET standards activity.

Due to its powerful architecture, SONET will integrate existing point-to-point fiber links into true networks that will route single voice-grade channels (DS0) through the network without the need for multiple stages of multiplexing and demultiplexing. The eventual role of SONET and fiber systems will be not only to replace the dedicated copper lines that are used in transmitting multiplexed digital signals (DS1, DS3, etc.) between two points, but simultaneously to provide greater functionality, standardized fiber interfaces, higher transmission speeds, and reduced operational labor requirements. With SONET, fiber, for the first time, will fully take its place as part of the international telecommunications network fabric. Only then will there be a flexible, controllable network with centralized network management that can support features such as bandwidth on demand.

Because of its capacity to manage large amounts of bandwidth, combined with its reliability, simplicity, and cost-effectiveness, SONET will, in a few years, totally displace the existing installed base of nonstandard fiber and electronic equipment. Already SONET rings are becoming the norm within the local loop. Since SONET is a transport technology, it does not necessarily displace emerging technologies such as frame relay, SMDS, FDDI, BISDN, and ATM, because they can and will be carried by the SONET network (Figure 9.11).

Synchronous Optical Networks

INTRODUCTION

SONET networks will embrace a mix of equipment from different suppliers. SONET will span the interconnect or midspan meet between different carriers and will bridge public and private networks. Features like direct multiplexing and grooming of DS0s, add/drop capabilities, and standard optical interfaces will simplify the design and management of broadband networks. Its equipment will form the tributaries, interoffice trunks, and metropolitan and suburban backbones of public and private broadband networks. SONET topologies include:

- Point-to-point networks that concentrate traffic from tributaries
- Drop/insert networks that use simplified multiplexing to eliminate the need for back-to-back multiplexers
- Ring networks with drop and insert capability that allow traffic to be sent to a central office, backhauled, and quickly restored in case of failures
- Star networks with a central hub and branches that extend to remote locations
- Hybrid networks that use a combination of the previous topologies

The variety of topologies, myriad of network elements, and amount of information being transported are introducing a new level of synchronization requirements to the public telephone network.

NETWORK SYNCHRONIZATION

Synchronization is necessary for all digital networks. The original, single United States synchronization system developed by AT&T was located in Hillsboro, Missouri. This PRC and its synchronization distribution infrastructure have

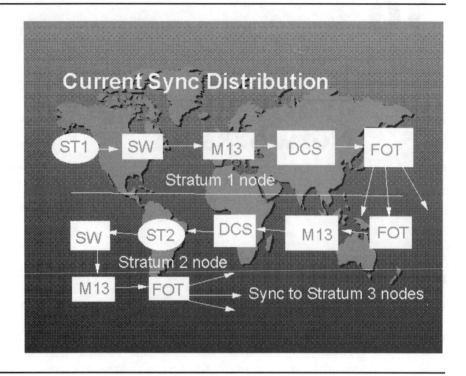

FIGURE 10.1   Current synchronous clock distribution.

since been replaced by 14 PRCs distributed throughout AT&T territory. Other carriers have similar arrangements. From a PRC, synchronization is passed through the network by asynchronous DS1 signals from master to slave clocks. These slave clocks then transfer timing to other slave clocks. Although this master-slave topology continues down the network to the end node, the synchronization is not usually passed directly from clock to clock. Instead, it traverses a complex chain of multiplexing, switching, cross-connection, and transmission systems (Figure 10.1). This clocking issue indicates the difficulty of maintaining accurate references over the long distances and intervening network elements within public telephone networks. To remedy this, the Bellcore technical advisory, TA-1244, specifies a new generation of clock systems. It describes a three-link system from the ST1 source; that is, a direct ST1 to ST2 to ST3 clock source (Figure 10.2). This timing is not affected by intervening clocks in a deeply layered network containing many tandem clock connections because there is direct access to an ST1 PRC. A local PRC is deployed within each network sector or at gateway locations.

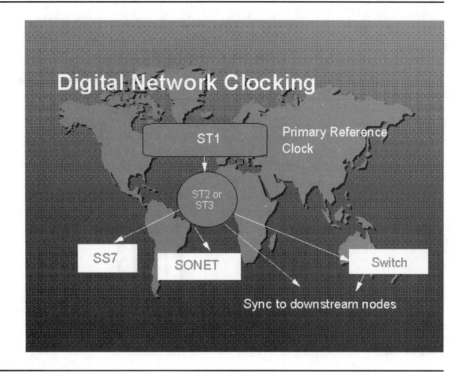

FIGURE 10.2   Digital network clocking.

SONET MULTIPLEXERS

A multiplexer allows:

- Sharing of a single high-speed/high-volume transmission line among several users
- Reductions in the number of transmission links by directing traffic to a higher bandwidth link
- Reduction in overall communications costs
- Configuration of diverse network topologies such as point-to-point, ring, tree, or hub

SONET multiplexers come in two varieties—a terminal concentrator type with many-to-one and one-to-many multiplexing/demultiplexing and a new type of ADM. Both types convert between electrical and optical signal environments. Internally, SONET multiplexers propagate STS-1 and STS-3 electrical

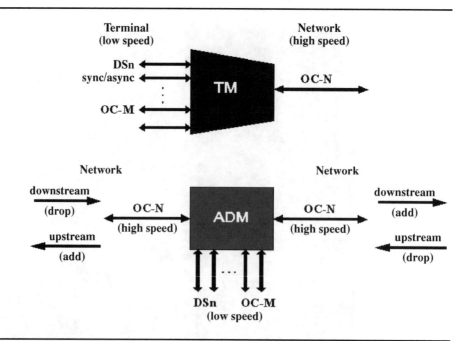

FIGURE 10.3   Conceptual difference between TM and ADM.

signals for DSn mapping, interconnecting, and distributing them at the local site, then multiplexes them to form higher-rate STS-n and OC-n signals. In an ADM multiplexer, the incoming and outgoing signals are at identical rates, while the lower-speed signals are dropped and added to it (Figure 10.3). The ADM allows lower-rate signals to be *groomed and filled*. That is, taken from incoming signals and added to outgoing higher-rate signals. Nonterminating signals pass through transparently from one port to another. Both TM and ADM multiplexers have a rich selection of interfaces and features that permit the configuration of simple as well as complex SONET networks.

Interfaces

SONET multiplexers are compatible with the existing telecommunication infrastructure because they support asynchronous as well as synchronous electrical interfaces for DS1, DS1c, DS2, and DS3 signals—and their mapping into the corresponding VT1.5, VT3, VT6, and STS-1 payloads. The DS1 is the most widely used interface and is available in almost all SONET multiplexers. High-speed optical OC-n signals are provisioned on a selective basis where $n = 1, 3, 9, 12, 18, 24, 36,$ and 48. The actual multiplexer data rate depends upon whether it

TABLE 10.1   Predominant rates for NEs vs. Serving Areas.

	multiplexers transport feeder		RFTs access only	WDCS transport only	BDCS transport only
inner city	OC–12 OC–24 OC–48	OC–3 OC–12 OC–24	OC–3 OC–12	OC–3 OC–12 OC–24 OC–48	OC–12 OC–24 OC–48
suburban	OC–3 OC–12	OC–3 OC–12	OC–1 OC–3	OC–3 OC–12	OC–12 OC–24
rural	OC–1 OC–3	OC–1 OC–3	OC–1 OC–3	OC–1 OC–3	OC–12

is deployed in a high or low traffic area (Table 10.1). In inner cities, where bulk transmission is crucial, OC-12, OC-24, and OC-48 rates dominate the network transport. Likewise, OC-3, OC-12, and OC-24 rates provide adequate capacity for feeder applications. In suburban areas, OC-3 and OC-12 are adequate to meet the traffic needs, while OC-1 and OC-3 are sufficient for rural areas.

Features

SONET multiplexers offer a wide selection of features:

- Direct termination of OC-n signals, where n = 1, 3, 9, 12, 18, 24, 36, and 48.
- Termination of low-speed, clear-channel DSn signals.
- Multiplexing/demultiplexing of DSn signals from/to OC-n ports. Multiplexers use a floating VT mapping structure for DS1, DS1c, and DS2, while DS3 signals are mapped directly into STS-1s.
- Add/drop of DSn signals from an OC-n pipe, which gives SONET networking its unique capability. The signals are either dropped when traveling in the downstream direction and added upstream or dropped from upstream signals and added downstream. Unaffected signals simply pass through the multiplexer.
- Remote and local operations interface to operations, administrations, maintenance, and provisioning. The EOC is the conduit for management commands.
- Additional features include VT add/drop, VT locked-mode operation, STS-1 add/drop, DS3 SYNTRAN mapping, ring applications with APS, loop-back capability for access, test, and in-service rolling.

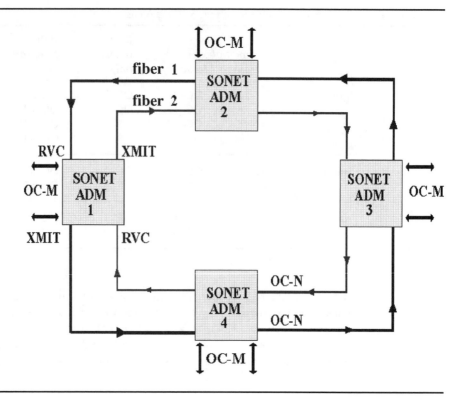

FIGURE 10.4   Dual ring/unidirectional structure.

Ring Applications

ADMs may be interconnected to form a ring (Figure 10.4). For *path protection switching* (PPS), two unidirectional fiber pairs are used between ADMs to back each other up in the event of a single fiber failure. In a dual-ring structure, two separate fiber rings are used. Each ADM on the ring transmits and receives simultaneously. Both fibers are monitored but only one is used by the ADM. Thus, a single fiber failure is serviced by switching to the second.

REMOTE FIBER TERMINAL (RFT)

The application of *remote fiber terminals* (RFTs) into the loop and distribution net–work has a long history in wire technology. Originally, miles of twisted copper pairs were bundled together to serve as access and feeder sections at the same time. One pair of copper wires was dedicated to voice or *plain old telephone service* (POTS) traffic, all the way from the customer's premises to the central office. In the 1960s, *analog loop carrier* (ALC) systems evolved to provide pair gain in the local loop or access network, while T1 digital technology became available for interoffice traffic. AT&T's SLC 96 systems in 1979 introduced digital technology to the loop. SLC 96 concentrates 24 POTS lines or channels over a single T1 line. Four T1 lines serve 96 subscribers and a fifth T1 line provides protection switching. The SLC 96 or *universal digital loop carriers* (UDLC) were never intended as long-term solutions. Because UDLC accommodates and interfaces with all types of switches—SPC, mechanical, analog, and digital—design compromises were made that limited performance. Nonetheless, the use of digital technology in the loop was shown to make economic sense and work was begun on more advanced versions.

During the 1980s, Bellcore defined an all-digital DS1-based loop system called *integrated digital loop carrier* (IDLC-TR303) to interface with a digital CO. In the late 1980s, a migration strategy to *fiber in the loop* (FITL) with SONET standards was defined by Bellcore as a supplement to the IDLC and TR303. Today, SONET has emerged as a physical layer alternative for local loop transport as well. All RDTs or RFTs are SONET-compatible. Deploying SONET in the local loop provides

- Consistency of interfaces and standards in the access section with those in the feeder and transport sections.
- Replacement of millions of copper cable miles with bandwidth-rich fiber. This will be expensive, but it will improve transmission and reliability and new services will result in significant revenues to offset the costs.

A remote terminal with a fiber interface provides a fully digital path from customer premises back to the central office. It supplies an end-to-end fiber path with a full SONET capacity. On business premises, the RFT offers both narrowband (64 Kb/s) and wideband (DS1-1.5 Mb/s to DS3-45 Mb/s) interfaces directly off the RFT cabinet (Figure 10.5). In this scenario, the entire loop from the business premises to the CO would be digital.

Due to its lower density, the residential network section today uses a different topology. Fiber is being extended from the RFTs to the *optical network units* (ONUs), with copper pairs going into the home. In the future, the copper pairs will be replaced with fiber as the demand for new broadband services increases.

DIGITAL CROSS-CONNECT (DCS)

Like multiplexers, digital cross-connects serve many purposes. They

- Terminate digital signals and facilities
- Cross-connect same-level signals automatically
- Cross-connect constituent (tributary) signals automatically
- Provide transparent grooming and routing
- Optimize use of network elements and facilities

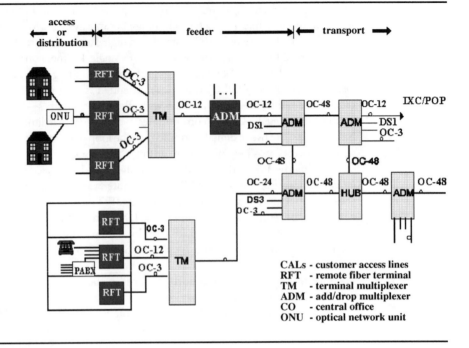

FIGURE 10.5   SONET network.

- Allow hubbing and network management capability
- Reduce provisioning and administrative costs

There are two types of SONET DCSs—a wideband type (WDCS) and a broadband type (BDCS). Wideband DCS is limited to DS1s/VT1.5s. In the cross-connect process, the wideband DCS terminates the incoming DS1, DS3, and OC-n signals and generates new frames for outgoing signals. The BDCSs perform DS3/STS-1 and STS3c to STS3c cross-connection by terminating both DS3 and OC-n signals. The cross-connect function is technically closer to switching than multiplexing. SONET brings a variety of signals to the CO and offers the opportunity to perform more and more low-level cross-connecting within the CO. With fewer higher-level signals to be cross-connected outside the CO, wideband as well as broadband functionality can be combined optimally in a single system (Figure 10.6).

Interfaces

SONET DCSs provide both asynchronous and synchronous electrical interfaces for DS1 and DS3 signals. The OC–n optical signals are provisioned on a

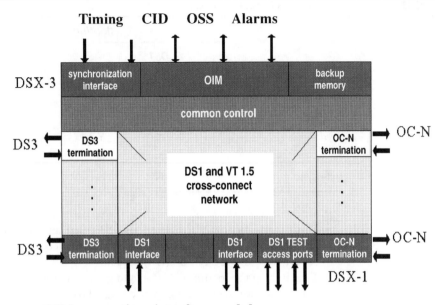

Timing   CID   OSS   Alarms

OIM - operations interface module
CID - craft I/F device
OSS - operations support system

FIGURE 10.6   SONET Cross-connect structure.

selective basis. For wideband cross-connect operations, OC-1, OC-3, OC-12, and OC-24 provide a reasonable mix of port sizes to combine hubbing, interoffice transport, and DS1/VT1.5 cross-connecting. Broadband DCS uses the higher rates offered by OC-12, OC-24, and OC-48, as well as STS-1 and STS-3 electrical signal. These are used for DS1/DS3 mapping and cross-connecting to VT1.5/STS-1 respectively, STS-3c/STS3c cross-connecting, and for intraoffice equipment interconnection.

DCS Features

Wideband cross–connect system functions include:

- Providing a low-speed DS1 clear-channel interface to cross-connect DS1 to DS1, DS3, or floating VT1.5. The transparent cross-connection includes a framing bit (F Bit), ABCD coding, CRC6, yellow AIS, and the ESF data link.

TABLE 10.2   Cross-connect signal matrix.

floating VT1.5	floating VT1.5				
	async DS3	DS1 CC	async	bit sync	byte sync
async DS3	Y	Y	Y	N	N
DS1 CC	Y	Y	Y	N	N
async	Y	Y	Y	N	N
bit sync	N	N	N	Y	N
byte sync	N	N	N	N	Y

- Terminating incoming asynchronous DS3 signals and generating new frames for outgoing signals as well as for cross-connecting constituent DS1s to different outgoing DS1, DS3, or VT1.5 signals.
- Terminating the incoming OC-n and generating new frames for outgoing signals and, at the same time, processing STS-1 overhead and frame alignment. Constituent VT1.5s are cross-connected to different outgoing signals (Table 10.2).
- Providing an interface for operations, administration, maintenance, and provisioning functions that include memory and database administration, performance monitoring, alarm surveillance, and signal facility maintenance with loop-back.
- Additional features may include enhanced DS1 performance monitoring, APS, in-service rolling, DS1 broadcast, side-door port for bypass, downloadable software generics, VT2, VT3, and VT6 cross-connection, and more.

Broadband cross-connect system functions include:

- Using DS3 clear-channel interfaces for both asynchronous and synchronous signals to provide cross-connection of DS3 to another DS3, STS-1 (OC-n), or itself for loop-back.
- Terminating an OC-n signal while constituent STS-1s and STS3cs perform cross-connect functions. An STS-1 may cross-connect to DS3s, to another STS-1 within the same OC-n, or to itself for loop-back. An STS-3c only cross-connects with STS-3c or to itself for loop-back function.
- Providing an operations interface for operations, administration, maintenance, and provisioning functions that is similar to those in WDCS.
- Additional features include broadcast capability (useful in video services), enhanced DS3 performance monitoring, APS, in-service rolling, downloadable software generics, and others.

SWITCHING

Some manufacturers have gone beyond simple, standalone SONET network elements and have integrated switches. Cross-connect functions may be integrated within the switch element or be an adjunct to the switch and only perform a cross-connect as part of the CO-hub (Figure 10.7). SONET technology within the switch is important because it

Meets the demands of customers for higher bandwidth and new dial-up services.

Automates operations, administration, maintenance, and provisioning processes, thereby reducing network engineering costs.

Gives network providers fully integrated network elements to build a platform that will allow them to offer new services rapidly and flexibly.

The marriage of SONET and switching systems will bring about a dramatic change in the wire center (Figure 10.8). SONET will simplify the network by replacing thousands of copper cables with a single-fiber cable system and it will eliminate redundant equipment, starting with the *main distribution frame* (MDF), *digital signal cross-connect frame* (DSX) patch panels, channel banks, standalone *fiber-optic terminals* (FOTs), M13 multiplexers, and standalone three/one/zero cross-connects.

ASYNCHRONOUS EQUIPMENT COMPATIBILITY

SONET network elements ensure compatibility with the existing telecommunications network. Interfaces with the current telecommunications components such as transmission system, outside plant fiber, operations systems, and power systems are part of the SONET standards. It is possible to interconnect SONET equipment at the standard hierarchical cross-connects such as DSX-1, DSX-1C, DSX-2, DSX-3, and DSX-4NA. In addition, SONET NEs provide electrical interfaces at STSX-1 and STSX-3 for SONET cross-connect facilities. It is, however, necessary to pay attention to the length of interconnecting cables; otherwise, timing problems and related alarms may occur (Table 10.3). The maximum cable distance between the terminal and the cross-connect frame should not exceed the lengths in the table.

SONET NEs also interact with the maintenance signals of the existing hierarchical rates (DS1, DS-1C, DS3, DS-4NA). With the use of VTs, SONET propagates the corresponding maintenance signals within the STS fabric and vice versa. In this way, the old DS maintenance signals and the new SONET maintenance signals are fully compatible. Fiber-distributing frames and optical DSXs connect any fiber cable to any regenerator without disassembling splices. The number of terminations and the type of fiber cable connections will further

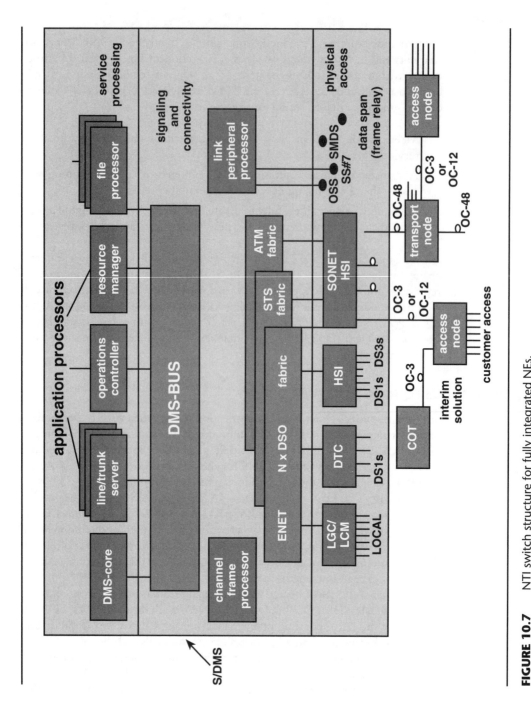

FIGURE 10.7   NTI switch structure for fully integrated NEs.

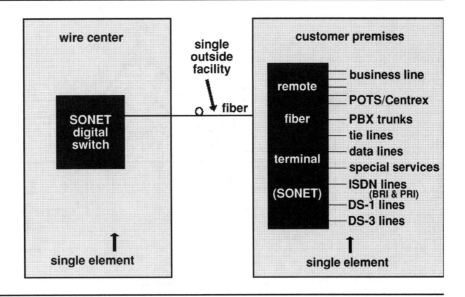

FIGURE 10.8   SONET-based access.

determine compatibility. SONET does not impose any new requirements for the mechanical interconnection of the fiber cables. The issue of mechanical compatibility for physical connections to SONET equipment is the same as that with the non–SONET fiber equipment.

APPLICATIONS

Today's asynchronous public telephone network uses a mix of manual patch panels, standalone digital cross–connect systems, and T1/T3 multiplexers to provide access and bandwidth management for new Telco/PTT services. The net-

TABLE 10.3   Maximum terminal and cross-connect frame distance (in feet)

cross-connect frame	maximum cable distance
DSX-1 and DSX-1C	655
DSX-2	1000
DSX-3 and STSX-1	450
DSX-4NA and STSX-3	225

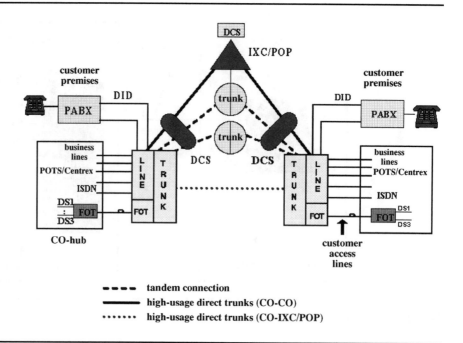

FIGURE 10.9   Today's networking scenario.

work elements are deployed at key locations to handle line and trunk grooming, cross-connecting, multiplexing, and demultiplexing. In Telco/PTT networks, CO-hubs consisting of manually adjusted MDFs, channel banks, customer access line blocks, standalone fiber terminals, trunking blocks, and so forth, along with local tandems and access tandems, are all interconnected (Figure 10.9). These network elements limit the capacity of the public telephone network for on-demand bandwidth services.

In contrast, SONET eliminates the need for external MDFs and DSX patch panels. It does away with channel banks, FOTs, and M13 multiplexers to provide more elegant and simpler networking solutions. SONET collapses the myriad of network elements into three primary elements: ADM/TM multiplexers, RFTs, and digital cross-connects. These are specialized to operate in the access or distribution, feeder, and transport sections of the public telephone network. SONET multiplexers (ADMs and TMs) are used in both the transport and feeder sections; RFTs make up the bulk of the access or distribution section. In residential areas, deployment of ONUs is already underway. The digital cross-connect systems that provide grooming and service expansion functions are part of the SONET CO-hub. These hubs primarily perform switching at rates that depend on the section of the network in which they are positioned.

Self-healing Ring Structures

Fiber rings are being used in the business districts of major cities in the United States (e.g., Cincinnati, Chicago, and New York, among others). Both local exchange carriers and alternate local transport companies are pursuing ring deployments vigorously. The use of automatic switching, which selects the secondary source of traffic in the event of a cable cut, makes them "self-healing" so that a single point of failure cannot stop two locations on the ring from communicating. This high survivability is critically important for communication links to business operations. Initially, the focus of survivable circuits was on the crucial, high-volume trunks that linked telephone company central offices with interexchange carrier points-of-presence. This was extended to alternate central office access links and beyond the central business districts, where self-healing fiber networks are now reaching the surrounding industrial parks and suburban business locations.

SONET technology facilitates the creation of fiber ring structures and greatly enhances their functionality and economical operation. Indeed, some Telco/PTTs have indicated that their interest in SONET is almost solely based on obtaining add/drop multiplexer capability that enables the insertion and extraction of traffic into survivable fiber rings.

Dual Ring/Unidirectional Structure

Ring structures that consist of two fibers—one as a working path and the other as its backup—are most suited for collecting, routing, transporting, and distributing traffic, especially with the add/drop functionality of SONET multiplexers (Figure 10.10). The total OC-N ring bandwidth is shared by all the nodes in the ring. A SONET ADM, DCS, or any other functionally equivalent network element may be employed, although in most ring applications the ADM provides the best fit. Both fiber cables carry the same traffic, but in opposite directions. One fiber is the working or active fiber, while the other is the backup. Each node in the ring simultaneously transmits, receives, and monitors the two opposing paths. During normal operation, the node selects the working fiber signal for STS/VT processing and handling. In the event of a fiber failure, a path-failed alarm signal is propagated downstream to all nodes. Since the downstream node monitors both fiber signals, it selects the path that has a valid signal, in this case, the signal from the backup fiber. In this way, the dual-ring structure offers a redundant signal that is identified and selected by individual nodes in the event of a single fiber failure. At least two consecutive failures of the fiber transmission path must occur to disrupt service to any customer location—an unlikely event.

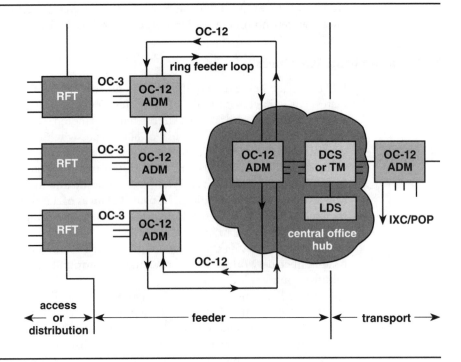

FIGURE 10.10   SONET suburban ring-based network.

Interoffice Ring

The SONET ring may interconnect central offices and provide a very reliable and flexible DS3/DS1 path between COs (Figure 10.11). The OC-N capacity of the ring accommodates the DS3/DS1 bandwidth needs of the three COs and the hub office. Since all the ring traffic is not processed at the hub, a smaller DCS may suffice to handle the outside traffic homing in on the hub office.

Inasmuch as all segments of the ring do not need the same capacity, one advantage of a SONET ring is that bandwidth capacity can be changed simply by adding higher- or lower-rate network elements. For additional bulk DS3 traffic between two COs, a separate but direct CO-to-CO fiber system can be established with its own line-protection switching. The ring then would be off-loaded and the available bandwidth used to support the traffic growth of other CO pairs.

Hierarchical Rings

Geographic considerations, as well as bandwidth management, may dictate that a network be constructed from multiple, interconnected rings. A multiring topology segregates traffic—an interoffice ring can carry STS traffic while an

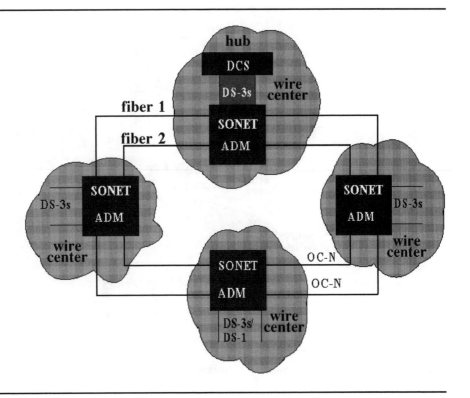

FIGURE 10.11   Interoffice ring applications.

access ring carries VT traffic. From the POP in Figure 10.12, STS-1 signals from an interexchange carrier are carried on the interoffice ring to the serving node and then terminated. DS3s are terminated by the SONET ADMs at the POP; from there they are transported within an STS-1 signal on the interoffice ring. The DS3s are processed by serving nodes that are attached to the interoffice ring. The demultiplexing of the STS-1 signals into VTs/DS1s and the routing of VT/DS1s to the access ring are accomplished at the primary and secondary serving nodes. The serving nodes in the access ring terminate the VT/DS1s as appropriate. Similarly, VT traffic from customers on the access ring is routed via the nodes onto the interoffice ring and then to the POP.

Both the interoffice and access rings function independently, with primary and secondary nodes coordinating operations between the two rings and satisfying the dual-ring, unidirectional transmission. Service is not affected by a primary or secondary node failure, nor by a simultaneous single failure in both rings. Furthermore, a failure in one ring has no effect on the operation of the

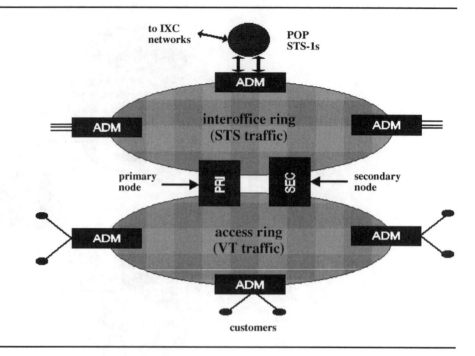

FIGURE 10.12   Hierarchical multi ring structure.

other. This robustness makes the multiring an excellent choice for use in interoffice as well as loop applications.

Local Telephone Loop

Every residence telephone is served by a local loop that is the actual physical line that connects the subscriber's telephone(s) to the switch at the CO. By lifting the telephone handset, the loop is closed, thereby completing an electrical circuit that carries the CO-generated dial tone. The DLC offers a way to gain more channel capacity without adding new wiring. The DLC multiplexes 24 voice conversations on a single pair of wires that formerly could only carry a single voice conversation. On copper wire, which is often what the loop is made of, this is usually at T-1 rate (1.544 Mb/s). Fiber-optic circuits offer even greater bandwidths, a necessity for emerging multimedia services. The optical loop carrier essentially substitutes optical fiber media for the older copper wire running between the central office and the remote terminal that serviced an area of several thousand homes.

For years telephone companies have been installing fiber-optic cable in

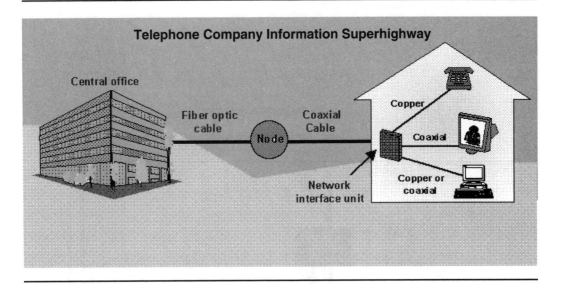

FIGURE 10.13   Telephone company information superhighway.

their networks. Long-distance or interexchange carriers were the first to do so, laying it for intercity trunks. Next, local exchange carriers used fiber to connect to IXC points-of-presence. Ultimately, the web was extended to central offices and then to digital loop carriers. Today, much of the telecommunication infrastructure is comprised of fiber. Telephone companies are now eyeing the local loop and trying to determine the best way to spin the final web to provide a high speed "pipe" capable of carrying high volumes of interactive voice, data, and video to homes and businesses (Figure 10.13).

The use of ring structures in the access loop or the network feeder section simplifies the process of traffic collection and distribution. As shown in Figure 10.14, three RFTs communicate with the local digital switch, using SONET ADMs interconnected in a ring. The SONET ADMs that collect and route traffic to the central office also pick up traffic that originates at the RFTs. The RFTs connect to the SONET ADM over an OC-3 or OC-12 fiber link, depending on customer service demands.

Cable TV Fiber Rings

Historically, each cable TV headend was an independent entity. Over the past decade with the introduction of fiber and digital video transport, fiber links have been established between a hierarchy of front ends (Figure 10.15). Cable TV

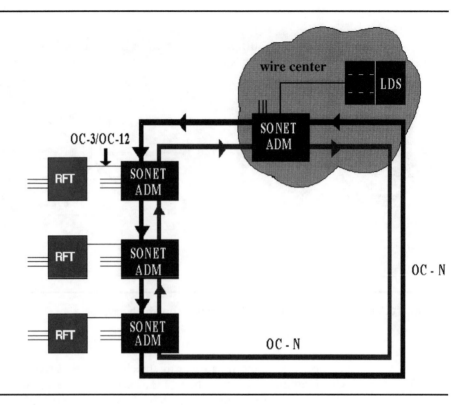

FIGURE 10.14 Loop (Feeder) ring application.

companies are using hybrid fiber/coax technologies in which copper cabling is overlaid with fiber-optic cable from the CO to the community or node, then a mixture of existing coaxial cable and copper is used to connect to the residence.

As cable operators deliver more advanced services and improve network reliability, they use ring topologies between these headends (Figure 10.16). Dual rings permit path redundancy because when network failures are detected the signal flow over the ring is reversed so that it arrives at the destination node by circling in the opposite direction. This is backed up by a second level of redundancy. Should the entire dedicated node ring fail, the transmissions are all switched over to the other ring. The first ring of the dual ring passes through all nodes to deliver traditional cable TV services and to carry the reverse signals for the nodes back to the headend. The second ring carries telephony and video service to all nodes and the relevant information back to the headend. Each ring consists of four fibers. Two fibers facilitate the deployment of the digital loop carrier, SONET add/drop multiplexers, and the other hardware needed to

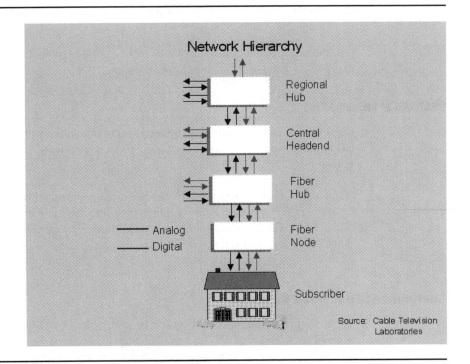

FIGURE 10.15 Cable TV network hierarchy.

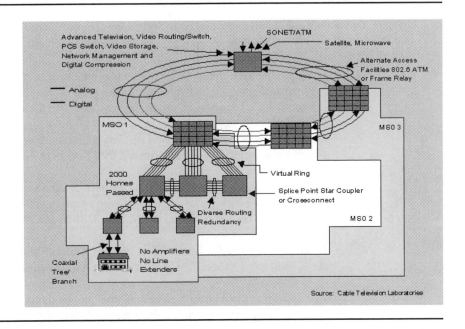

FIGURE 10.16 CATV hub network architecture.

multiplex subscriber communications paths onto a high-speed signal. The second pair of fibers serves as a redundant path.

RESEARCH NETWORKS

Gigabit per second networks are being tested in both government and commercial applications. There are three main research thrusts in the gigabit test bed initiatives:

1. To gain an understanding of the technologies needed to support gigabit networks, particularly for switching, transmission, and protocols.
2. To discover applications that are made feasible by gigabit networks.
3. To develop computer operating systems that successfully accept data at 622 Mb/s and higher rates.

Government-Sponsored Research

There are five separate networking test beds supported by the *Corporation for National Research Initiatives* (CNRI), the *National Science Foundation* (NSF), DARPA, and participating long-distance carriers, regional telephone companies, and switch vendors. The test beds all use SONET transmission backbones at 2.488 Gb/s and 622 Mb/s channels. Each of the five projects focuses on a different aspect of high-speed networking and computing:

- *Multimedia*—The Aurora project compares the performance of fixed-length packets and variable-length packets for the transport and switching of multimedia information.
- *High-speed network control*—The Blanca project connects two networks over a transcontinental T3 line. Research is focused on real-time protocols for the control and transport of remote images, as well as their visualization and simulation.
- *Distributed computer computation*—The Casa project connects *high-performance parallel interface* (HIPPI) directly over SONET, with the intention of investigating parallel computing.
- *High-speed data delivery*—The Nectar project examines the impact of the full 2.488 Gb/s bandwidth on a single end-point. A general gateway from a local area HIPPI network to ATM over a 30 km SONET link is being developed.
- *Distributed high-speed networks*—The Vistanet project links a HIPPI LAN to the ATM over a SONET WAN. A major focus is medical imaging.

The best known of these initiatives is the Aurora gigabit test bed which has been operational since May 1993; although it was not until 1994 that the first data flowed over the network. The fiber network interconnects researchers at IBM, Bellcore, the Massachusetts Institute of Technology (MIT), and the University of Pennsylvania. Throughout the test bed is SONET gear. The backbone is a 2.5 Gb/s OC-48 link that is demultiplexed to OC-12 (622 Mb/s) links at or nearby each of the four end-points. Two switching technologies are being evaluated: One is ATM, which uses fixed-length cells and the other is *packet transfer mode* (PTM), which uses variable-length packets. PTM is supported by IBM's PRIZMA switch (originally named PlaNET) and ORBIT ring. The current implementation of PRIZMA is an eight-by-eight switch, with each port running at a speed of 1 Gb/s.

Telephony Company Gigabit Tests

Long-distance and local carriers are operating test bed networks to examine the interaction between ATM and SONET services. Sprint, for example, has in-

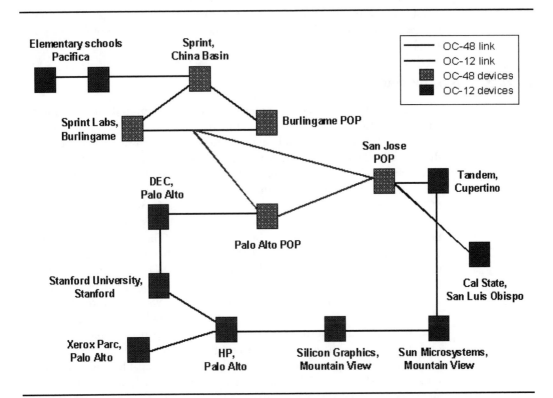

FIGURE 10.17   Sprint SONET ATM test network.

vested millions of dollars to create the Silicon Valley Test Track that connects six computer vendors to Sprint's Advanced Technologies Laboratory (Figure 10.17). The network uses two fiber-optic rings: one at 622 Mb/s and the other at 2.5 Gb/s. Workstation vendors are building computers that can accept 155 Mb/s SONET pipes with ATM over them. They are using the network to test the performance of their product when they are connected to high-speed circuits and to develop applications that take advantage of the network's bandwidth.

Medical Test Bed

The Johns Hopkins Center for Information-Enhanced Medicine has constructed a pilot ATM research network to tie together the Johns Hopkins School of Medicine and the University of Maryland School of Medicine (UMMC). The shared data includes critical images that are an important factor in reconstructing the facial deformities and skull fractures of children. The network, which features a medical image database and an archiving system, allows UMMC to send detailed CAT scans and MRIs to Johns Hopkins for research. Researchers at John Hopkins can then quickly access the UMMC radiology department's scanned images and UMMC can collect any of the images housed on Johns Hopkins' workstations.

BYPASS AND DARK FIBER

Recognizing the considerable monetary investment required for replacing the embedded base of asynchronous equipment, the telephone companies will be slow to replace perfectly usable existing equipment. At issue is when to begin the upgrade. The North American and European digital transmission rates fall below those of their SONET counterparts. Although SONET NEs support the major existing framing formats, the reverse is not true. It is not possible to cram SONET signals into the existing network. Pressure from private networks will act as a catalyst for SONET deployment. Within the confines of the corporation, the seeds of SONET have begun to sprout. They could force the telephone companies into deploying SONET earlier than planned. In cases where the telephone company is unwilling to collocate SONET equipment at the CO, corporations are choosing either dark fiber routing or bypass providers.

The issue of dark fiber dates back to the mid 1980s, when the local telephone companies installed it as the foundation for high-speed, point-to-point links in private user networks. The telephone companies supplied the fiber-optic lines. Today, the electronics that originally were attached for a DS0 in the 1980s could be replaced with SONET gear that is capable of supporting an OC-3 (155 Mb/s) line. The telephone companies initially offered the fiber on an individual case basis. That is, they worked out a method with users to cover the cost of purchasing, installing, and maintaining the fiber. But in the 1990s, when the FCC decided that formal tariffs were required for the service, and the user demand was increasing, the carriers tried to exit. As time passes, there will be situations where the telephone companies will have to take the initiative because the service and bandwidth requirements can only be satisfied with SONET NEs and also because new installations will favor it.

The positioning of SONET must begin in a network environment dominated by asynchronous multiplexing of the DS N hierarchy. Because existing asynchronous multiplexers and SONET NEs can coexist in the same network, there will be no immediate wide-scale replacement of DS 3 networks, but rather a gradual overlay by SONET equipment. Consequently, any general application will need to be studied from three perspectives: the existing network solution, the temporary hybrid solution, and the pure SONET solution.

CONCLUSION

SONET has the promise of bringing public and private networks together with a common equipment and control format that makes sense in both environments. It provides a variety of network topologies to select from. Which of these topologies will be the most desirable will depend on the speed of the SONET migration into the public network. The major inroad of SONET equipment to date is in the local loop where a number of providers—telephone and cable TV—are vying to provide information and entertainment services. Many of these pioneer companies are operating gigabit test beds as well as testing the market for broadband services. Progress in the long-haul telephone network is proceeding more slowly. Although SONET supports the current asynchronous signals, the reverse is not true, necessitating a transition period in which isolated SONET network islands will be used along with current asynchronous long-haul networks until pure SONET implementations become the norm.

STRATEGIC CONSIDERATIONS

Planning Considerations

INTRODUCTION

The fundamental changes taking place in telecommunications with regard to technology and business orientation are creating new business realities. Businesses can no longer stand alone—all are interconnected by a vast global communications network. How effectively this network is used may very well determine whether a particular enterprise survives or dies.

The Telco/PTTs are making available to their customers sophisticated broadband switching fabrics and services, as well as new methods of managing them. The evolving broadband public network will serve business and residential customers with equal facility. All users will eventually use the same basic communications technologies for an array of bandwidth-hungry applications. Now is the time to plan, before the enterprise network reels under the impact of high-bandwidth applications—applications that have become as much a part of the desktop during the 1990s as personal computers were in the 1980s. Businesses need to devise ways to gain competitive advantages through the intelligent application of their telecommunications networks.

PLANNING FOR THE FUTURE

Network managers know that their networks will be vastly different in several years, but what these changes will be remains uncertain. The overriding responsibility for the manager is to meet the immediate business needs of his or her company and still justify the expenses of network enhancements. In planning a network, a strategic approach must be used that builds toward the broadband future while still addressing short-term networking needs. The ideal approach deploys broadband technologies as they are proven without disrupting existing applications and services. Change begins with a complete analysis that matches the business needs and expectations with vendor and carrier offerings.

Reengineering the Corporate Network

Although the gap between LANs and WANs has narrowed, each has its own set of rules. The LAN is often considered a combination of cabling, hardware, and software, all used for connecting computers. Today, this view must be extended to include a web of communications links, both within and between locations, that transports voice and data. Accommodating the bandwidth needs of advanced business transactions is crucial to taking advantage of this communications network. The infrastructure decisions for both public and private networks remain tied to the bandwidth demands of everyday business transactions. The anticipated progression of broadband technology from ATM LAN switching to desktop applications and finally to long-distance SONET/SDH networks, will change the networking equation (Figure 11.1).

Selecting among Broadband Services

The impact that an application has upon a communication network depends heavily on the line speed and the amount of data to be transferred during each connection (Figure 11.2). The networking alternatives for handling a particular application and the accompanying questions can engulf the planner. How much information will be transferred? Should a private network or carrier services be used?

Examining the way networks are priced can provide some answers. For

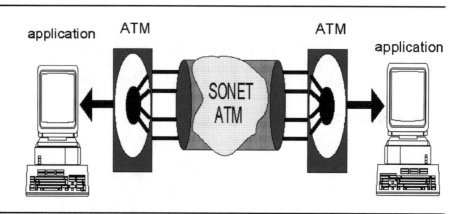

FIGURE 11.1   Desktop-to-desktop broadband network.

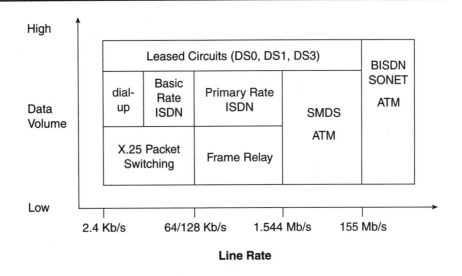

FIGURE 11.2   Service selection criteria.

example, the tariffs for switched data services such as X.25 or frame relay are based on the size of the data transfer, while the cost of using an ISDN network depends only on the call duration. Switched data technologies such as SMDS and ATM share the transmission network among several simultaneous connections, thereby lowering communications costs. In contrast, ISDN temporarily monopolizes the network for the call duration. Finally, leased lines provide the greatest amount of bandwidth and security for applications, but they are usually the most costly solution.

Short-term Migration Plan

Although there is never an optimal network architecture, there are some guidelines for designing or upgrading a network. As unexpected breakthroughs in communications technology are unlikely in the near term, communications managers should go ahead and modernize their networks. The demand for bandwidth will continue to increase and competitive pressures will continue to push organizations toward networks that are able to adapt as business goals change or as new technologies evolve.

Network modernization should begin with a five-year plan that does not commit itself to untested approaches that might cause problems. Most organizations should implement 10 Mb/s token ring or switched Ethernet on every desktop. For networked multimedia, 10 Mb/s is more than adequate for most users, especially with the MPEG compressed video standard. Nonetheless, bandwidth-intensive users may need 100 Mb/s fast Ethernet or 155 Mb/s ATM. Backbones should migrate to ATM and WANs should use virtual private networks for voice and frame relay or SONET for data. Managers should be cautious with FDDI, which appears to be losing vendor support, and with ATM to the desktop, which is too powerful for most applications. With this type of technology menu in hand, managers can make clear choices.

Access Links

The considerations for access links between a company's premise equipment and a carrier's switching facility should include: switched-56 Kb/s, ISDN BRI, 56 Kb/s *digital dataphone service* (DDS), T1, integrated T1, and ISDN *primary rate interface* (PRI).

Cabling

When considering a cable upgrade, it should be kept in mind that the dominant cost is not for the cable, but for the cost of installation. Do not skimp on LAN cabling. Consider using:

- Separate twisted-pair pulled to each workstation and to each telephone with at least two pairs per phone (depending on PBX requirements) and four pairs per workstation
- Multimode fiber or Category 5 as the backbone between floors within the building
- Single-mode or multimode fiber running across a campus or in cases where distances exceed two kilometers

LAN Migration

Some LAN migration guidelines are:

- As traffic builds, networks should migrate from shared-media LANs to switched LAN and fast Ethernet, and eventually to ATM.
- Ethernet or token ring should be connected to every desktop. For most users, 10 Mb/s will suffice. Fast Ethernet at 100 Mb/s or ATM at 155 Mb/s to each desktop may be needed for the most bandwidth-hungry applications. One can use twisted-pair to keep the cost down, but managers should seriously consider using fiber for eventual growth into ATM or to overcome distance limitations.
- Deploy ATM backbones when they become practical.

A number of progressive organizations are installing ATM-based LANs to gain the economies and efficiencies of increased backbone network bandwidth.

High-speed routers and bridges interconnect the LANs. These organizations will be ready for the second generation of routers, multiprocessor *reduced instruction set computing* (RISC) machines, which will support very high-speed routing. They will connect directly to the ATM infrastructure, providing a migration path to the desktop where ATM will enable the use of more sophisticated groupware, video, voice, and multimedia operations.

WAN Migration

Until multimedia applications become more common, most of today's applications will continue to consume less than 128 Kb/s per information transfer, well below the threshold capacity of megabit fiber-optic networks. To be able to cost-justify a broadband network for such applications, they must be combined over a common facility. Total bandwidth is the key network parameter; planners must audit the company private lines and tally the switched public network calls to determine its extent. Once the broadband network is justified, some WAN migration guidelines may be applied.

- The cost of fiber-optic trunks, not including the multiplexing equipment, is approximately the same for T-carrier (T1 and T3) and SONET OC-1 and OC-3 services.
- Multiplex sources of information of various types onto T1 or T3 trunks.
- Use data equipment with high-speed interfaces such as DS1 or DS3 rather than serial interfaces.
- Consider using frame relay for data and virtual private networks for voice.
- Use SONET rings for metropolitan area networks, but not for long-haul networks where the installations of SONET repeaters represent a considerable expense.
- When available, deploy ATM for integrated voice/data applications.

Finally, the promise of ATM as a global area network strategy should not be overlooked. The use of ATM in public and private networks will, for the first time, ensure seamless end-to-end connectivity. The global use of ATM depends on the availability of fiber-optic facilities. The rate at which SONET/SDH equipment is deployed within the Telco/PTT infrastructure will determine the rate at which high-speed, wide-area digital transport facilities become available. Fortunately, the groundwork for establishing critical mass is near completion.

LONG-TERM MIGRATION PLAN

A hybrid strategy, in which current data technologies coexist with ATM and SONET/SDH, is a realistic approach. It will protect the installed investment, keeping the token ring, Ethernet, and FDDI LANs intact while ATM adapters are being added in the hub or at the desktop.

Consider Carrier Services

By evaluating an enterprise-wide broadband network and comparing it with alternative networks, a potential customer will get a better sense of what broadband applications can and cannot deliver. Broadband networks will create a host of exciting services, including

Business video conferences

High-definition image transfer

Distance learning

Interactive TV

Real-time manufacture and performance data reviews

Ongoing market reports

Medical diagnosis and consultation

Working and shopping from home

Video-on-demand entertainment

Worldwide video telephone calls

To ensure that broadband networks merit business use, planners must evaluate available services, select those that meet the applications, and specify devices that provide flexibility and protect investment.

Weighing Price and Performance

There are certain economies that carrier services offer. Frame relay, where available from such carriers as AT&T, MCI, Sprint, and Wiltel, is usually a lot less expensive than leased lines because customers do not have to build a full mesh network between all of the points to be internetworked. SMDS, which is offered by local telephone carriers, is an economic alternative to metropolitan area networks. Other alternatives for WAN transport include existing leased circuits to carry compressed video, SNA, and LAN-to-LAN traffic; and carriers' virtual network services to carry voice traffic.

When evaluating the effective cost of these services, there are several aspects to consider

- *Fixed costs* are the one-time installation charges for circuits and equipment.
- *Periodic costs* are the recurring charges for leasing the circuits and can vary greatly according to distance and type of service and carrier.
- *Usage costs* are incurred each time the circuit is used and may be per minute, per packet, or per session.
- *Maintenance costs* are associated with equipment and network. For example,

some frame relay carriers offer end-to-end service that includes routers on customers' premises.

However, all things considered, the assessment of network technologies should be based upon more than just cost. The coalescing of the communications and computer industries is accelerating the switching and distribution capabilities that will provide the residential and business customer with powerful broadband services via the public-switched network. The challenge now facing businesses today is how to incorporate these services into on-demand networking solutions.

Internetworking LANs

One cost-effective migration strategy is to speed up LANs and provide switching capability. This gives full bandwidth to each user and may be implemented either with available Ethernet switches or with ATM. For many desktop users, 10 Mb/s is still adequate, and most hub vendors rely on some form of Ethernet switching. Since Ethernet switches are only one step along the path to ATM, wiring issues must be addressed today. If new wiring is being installed, or if existing wiring is being upgraded, it is important to ensure that the cable plant can support ATM. A rule of thumb is that for campus networks, fiber-optic cable should be used; in buildings, use class 5 unshielded twisted-pair as a minimum standard.

Advancements in lasers, switching, and broadband standards are hastening the use of ATM, the first communications technology equally at home in the LAN and the WAN. Today, many major data communications manufacturers have announced ATM products. ATM-based LAN switches are available and some private network installations of ATM are underway (Figure 11.3). For the first time, there is a logical way to integrate local and wide-area communications. ATM defines an open, interoperable interface between networks, permitting users to take advantage of high-performance features and services. Initially, ATM will advance in organizations that want to deploy bandwidth-hungry applications in a LAN or campus environment. Campus network installations based on ATM will spur growth in the ATM interconnection market as ATM forms the basis for emerging broadband-tariffed services over public networks. Within two years, a wide variety of ATM products will exist, including some from public network carriers. Frame relay and SMDS will be supported over ATM networks. The LAN-WAN barrier will begin to break down as similar transport protocols are used over both types of networks.

Global Ubiquity

The success of a networking technology is ultimately determined by its level of acceptance and its ability to solve user problems. The emerging high-speed networking technologies and services previously discussed are not yet globally

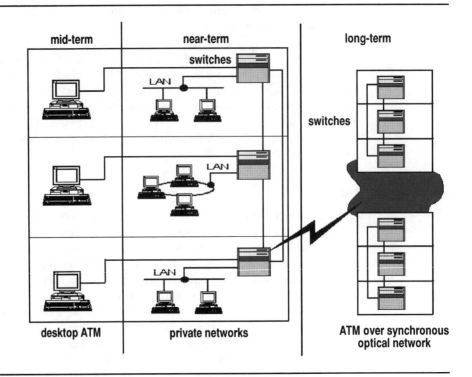

FIGURE 11.3   ATM progression.

available and will most likely change somewhat before they are universally accepted. When considering which technology to employ, users should understand how each fits into private and public networks, as well as whether it fits the access, transmission, or switching portions of the network. In the competitive race, all those portions will survive in some form.

It is anticipated that ATM will emerge as the dominant switching technology and SONET/SDH as the transport technology. Frame relay will be used as an access link to private and public networks. SMDS will provide the link to the public network for fiber LANs such as FDDI, as well as for the lower-speed Ethernet and token ring LANs. The migration has already begun; the expansion of existing fiber-optic installations and improvements in the reliability of high-speed transmissions over copper will lead to widespread ATM installation in the public telephone network. Many of the Telco/PTTs are developing ATM networks that will eventually interconnect—a task made possible by the existence of common standards for ATM in the United States, Europe, and Asia.

Telco/PTTs will offer transport services to the general public that will deliver advanced capabilities such as multiple channels of high-speed animation and data to the business market. Fortunately, network managers can make the necessary investments today to meet their imminent network challenges, while at the same time maintaining a clearly defined path. Consider the following:

- LANs and attached personal computers will continue to grab market share from earlier mainframe and minicomputer networks.
- X.25 still dominates data communications in terms of installed interfaces and actual use. However, frame relay is making in-roads into existing X.25 applications. Frame relay grew rapidly in the early 1990s into an international standard supported by many carriers and vendors. It is essentially the next generation of X.25 that uses high-grade digital lines to offer much greater speed.
- Both frame relay and ATM are suited to the access section of private and public networks. Frame relay addresses today's megabit transport and access. ATM addresses multimegabit transport and is applicable to switching as well.
- ATM is the core technology for broadband ISDN services.
- SONET/SDH will become the dominant public network transport of the next decade, beginning now to overlay and then replace asynchronous networks.

MANAGEMENT STANDARDS

Although proprietary networks were dominant for decades, the 1990s witnessed the decline of their influence. Users needed access to the latest productivity-enhancing technology, no matter who the supplier was. Proprietary networks locked users out of the decision-making process, forcing them to depend on single vendors. The rise of personal computers and LANs and the standardization of WANs have provided the opportunity for open, standard implementations. In contrast with the ease of its implementation, the network management of this equipment has been particularly difficult to standardize because there are not one, but two open standard protocol suites: the OSI reference model and the TCP/IP.

TCP/IP protocols are readily available and the technologically more advanced OSI protocols are not being developed quickly enough to supplant them. Although there was much discussion about the CMIP of OSI replacing the SNMP of TCP/IP, the standard for CMIP was not accepted before the SNMP evolved. As a result, SNMP II was released to fill many SNMP gaps, including the security issues raised by the government and private industries. Originally developed to manage the routes on the Internet, SNMP now accommodates virtually any manageable network device. Since many vendors have written SNMP interfaces for their products, that protocol has become the most widely accepted standard for network management.

NEW ROLES

For leading-edge corporations, networks are the predominant business platform, and the people concerned with their design and maintenance—from the support technician to the chief information officer—are the agents of change. Today the corporate network can be a business unit that is expected to contribute to the bottom line just like any other business profit organization in the company. And the user is a customer who should be treated just like any other customer. The information provided by the network is so important to an enterprise that networking people with management skills and knowledge are highly valued. What are these types of management skills and knowledge? They are the same as those used by other top executives, overlayed by *an awareness of changing communications technology*. That is, leadership, a strong sense of business practices, attention to the bottom line, an understanding of the corporate mission, and enough technological savvy to be the technology *expert* in the company.

CONCLUSION

Underestimating the amount of traffic on the enterprise network and the issues that arise in moving it along to remote locations, are not unusual in these, the early days of broadband LANs and WANs. Traditional network planners are not familiar with this environment. Fortunately, broadband networks are beginning to fulfill the promise of the information age. This revolution in data communications has provided a whole new generation of computer hardware and software applications that will be combined over broadband networks in ways that eventually will replace traditional computers, telephones, and televisions. The human and technological interfaces of multimedia communications are past the laboratory stage and have entered the mainstream, making businesses more productive. This process of making information more accessible, understandable, and usable is bandwidth-intensive and will provide further incentive for the rapid deployment of broadband networks.

CHAPTER
12

Digital Convergence

INTRODUCTION

Today, the way information is transmitted and the form in which the transfer occurs is fundamentally changing global communication networks. Facsimile machines in Vietnam, cellular telephones in Argentina, satellite dishes in Russia, and video terminals in Los Angeles are all part of a communications revolution that is upsetting the rules for where businesses reside and the way they operate. With a worldwide public communications infrastructure of high-capacity fiber cables, digital switches, and satellites, a tiny local firm and a giant multinational corporation can conduct business from any geographic location, 24 hours a day. Information that was once the domain of a few powerful corporations, now freely resides on the Internet and is accessible to anyone from his or her business or residence.

The resulting free-for-all for information services is redrawing the map of the cable TV, broadcast television, publishing, entertainment, telecommunications, and computer industries. This is occurring at a time when advances in digital electronics are propelling the convergence of the underlying technology engines that power these commercial enterprises. For an increasing number of businesses, this coming together of technologies known as multimedia is the most important advancement of the decade. (Multimedia brings text, image, sound, and video into the education and entertainment process, challenging all of the senses. Key components of multimedia applications that are in the forefront of change are groupware, electronic mail, and database management systems such as client/server.) The number of multimedia applications deriving from the ubiquitous PC is growing rapidly. In fact, the demands of these bandwidth-intensive applications have already noticeably impacted private as well as public networks. More than just the expansion of the bandwidth capacity of a network, a wholesale use of advanced communications to reengineer entire industries is opening up unprecedented new market opportunities while lowering the barriers of entry to existing markets.

The combined total value of these markets is staggering, exceeding trillions of dollars. The megamergers and partnering of some of the world's biggest corporations have come about because of the size of the service opportunities alone. Some companies are looking at new forms of entertainment such as movies on-demand and are seeking to take control of portions of the entertainment industry. Others view telecommunications, cable TV, and computers as fruitful areas for expansion. To serve this massive market, industries are evolving and, to a degree, coalescing. New competitors aided by the convergence phenomena are going after established companies and, with superior service, winning their customers away.

In the United States, competition from alternative providers and a constant demand for better service by customers are persuading national carriers to give up the monopoly protection that they have enjoyed for a century so that they might compete in the emerging markets. New services are shaping the broadband communications networks of the future, particularly in the last mile or local loop.

SERVICES

The future promises powerful, ultra high-speed networks both within companies and across the public information superhighways. Wired and wireless ramps will bring information to consumers no matter where they are located. Eventually these networks will shift the computer industry away from the desktop, the telecommunications industry away from voice, and the cable TV industry away from one-way programming, and will transform them all into a single coherent multimedia information fabric. Businesses already are setting the stage by assuming control of the access to the networks as well as the content and services that are offered. All types of devices—PCs, PDAs, digital televisions, and so forth—will tap into the broadband networks, accessing multimedia information, and connecting users with a multitude of on-line services.

The basic question is what are the actual benefits of the converged services that will flow across these networks? A company's view depends on its individual background. Cable TV, computing, and telecommunications giants are betting enormous amounts of money and resources on their belief that people will want interactive sports, movies on request, work at home, and endless access to caches of information. To create these services, it is important that the correct digital technology be used. The universal public telephone service is successful because there is a single telephone system for both business and residence. Wherever voice conversations originate, the communication system is still formed from the same basic technology. What differs is the type of usage.

The same is true for multimedia systems; converged networks that support interactive video, automated information agents, electronic shopping, and so forth also do not distinguish between locations (urban, rural, mobile, static, etc.). Systems architects must factor in the services that the networks support and the broad range of user motivations for, let us say, information versus entertainment services. There are differences between information usage and entertainment consumption. The consumers of each type of service are often the same people, yet their reasons for one over the other are different and the media needs are not the same. Finally, the appliances that people will use to access the services need to be considered: The PC is an easy choice because it is already interactive, but television has better visual connotations.

CONVERGENCE VEHICLES

Some companies believe that a television equipped with an intelligent cable box will be the primary interface to the information highway. Others feel that the convergence of computing and telephony revolves around PC-like appliances.

PC Centric

The digital superhighway will rely on the continued evolution of the PC. The design of the PC, which has remained fairly stable since its introduction in the 1980s, is now rapidly changing into a multimedia information appliance. An ongoing merging of the microprocessor with television technology is linking PCs with combinations of broadcast television, cable TV, and telephone networks. The PC already acts as a smart terminal for information services, a place to transmit and receive faxes, the portal to E-mail, and the creation station for business documents. Vendors offer computer telephony packages that permit users to do everything from routing calls, integrating telephone, fax, and E-mail functions, to connecting calls with database programs. Softphone packages provide teleconferencing, call transfer, speed-dialing, and other telephone-related support via the Windows interface.

A new generation of personal communication services are accessed by hand-held appliances based upon the familiar PC architecture, to offer universal connectivity to a variety of information (Figure 12.1). This merger of PC technology with telecommunications promises a seamless environment in which business and recreation goals are merged and met in the most simple and painless way for the user.

Although there is much to be done before the average PC becomes the optimal gateway to multimedia information, the industry has already begun to

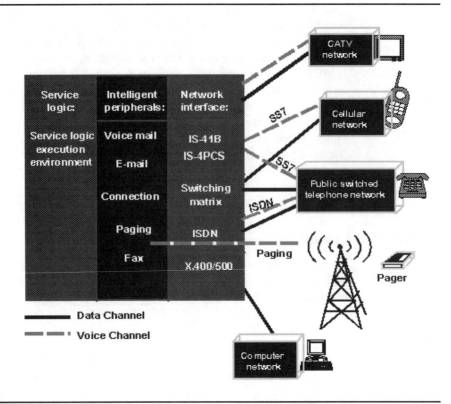

FIGURE 12.1   Universal multimedia access network.

include improved cable television and telephone interfaces. This converged PC model requires full 64-bit microprocessor capabilities that are controlled by a multitasking operating system, a fast internal bus, fast video, and vast amounts of storage. It also needs built-in software to interface with both information and entertainment networks.

Multimedia Information

The reach of multimedia services is extending to the individual consumer. Forms of wireless personal communicators in combination with fiber optic cabling to the residence and business are speeding the process. New multimedia technologies are blurring the traditional lines between computers and telephones. Regardless of the nature of the network—wireless, terrestrial, cable TV, competitive access provider, or interexchange carrier—multimedia services promote the effective processing and assimilation of information by people rather than computers. Providing new ways to communicate and relate with a variety of information resources as well as speeding the assimilation process, multimedia is creating changes in the dynamics in the workplace and often in the corporate culture itself. Understanding this complex field is crucial to the health of many industries that include telecommunications, cable, computer, and publishing, among others.

For multimedia companies whose expertise ranges from computers, consumer electronics, movie studios, television networks, and telecommunications, to publishing and cable television, the stakes are high. For them, the multimedia combination of text, images, sound, animation, and full-motion video promises to transform the base of hundreds of millions of personal computers into desktop workspaces, and the even larger numbers of televisions into interactive entertainment vehicles. Some mass media providers have initiated interactive public access to multimedia information; cable TV, pay-per-view, and home shopping channels are experimenting as they explore ways to provide two-way communications so that viewers can watch programs according to their personal schedules, provide feedback, play along with game shows, and order products and services on-line. Telephone companies, whose public networks already permit two-way communications also anticipate being able to store and transmit interactive services such as video-on-demand. The result of all this activity should be a dazzling array of new multimedia services for the consumer.

Delivery Systems

The components of these information highways are the public telephone and cable TV networks which have the broadband capacity to transport bandwidth-intensive files; appliances with the visual and audio capabilities to service multimedia applications; and the comprehensive system design that makes it all possible. For the user, immediate access to vital records is provided desktop-to-desktop via the information appliance, the display, and the network. The global desktop-to-desktop delivery systems (i.e., the public telephone networks) are undergoing a digitization that will support the transport of bandwidth-intensive multimedia files.

The implementation of more intelligent, higher bandwidth, massive storage capacity networks allows people to interact with machines as readily as they do with each other. A variety of interactive multimedia services—voice, video, interactive TV, image, and data services—have begun to be delivered to homes and businesses (Figure 12.2). They are or will be transmitted over fiber-optic cable, hybrid fiber or coaxial cable, coaxial cable, and copper wire. Fiber-optic cable has become the single most important conduit with much of it being installed in residential neighborhoods. Broadband trials have been underway for several years by telephone, cable TV, and computer companies, using SONET transport and ATM switching in their information delivery systems.

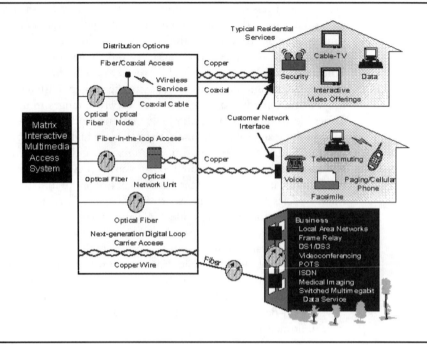

FIGURE 12.2   General multimedia delivery system.

To take advantage of these technological innovations, corporate managers need to consult with one of the network providers in order to understand and plan how their own systems will plug into the emerging information highways. The solution may be as simple as acquiring and managing a voice telephone network. Perhaps it may mean working with a local cable company to move enhanced telephone and video service into the local network. Or it could mean working with both the local telephone and cable company to integrate the local network into some type of high-speed switched network.

SERVICE PROVIDERS

Cable TV and telephone companies strongly believe that consumers will want new interactive and pay-per-view services (Table 12.1).

The successful deployment of multimedia services will merge the talents of companies that up until now have been separate and distinct. Content will be provided by today's information publishers and entertainment providers, transmission by traditional telephony and cable TV companies, and the actual appliance by computer software and hardware vendors (Figure 12.3).

TABLE 12.1   U.S. Market Size for Video Dialtone and Video-on-Demand

Year	Market Size (Billions $)	Growth (%)
1994	.2	—
1995	.3	50
1996	.5	66
1997	.8	60
1998	1.1	38
1999	1.2	9
2000	1.3	8

Source: IGI Consulting 1994

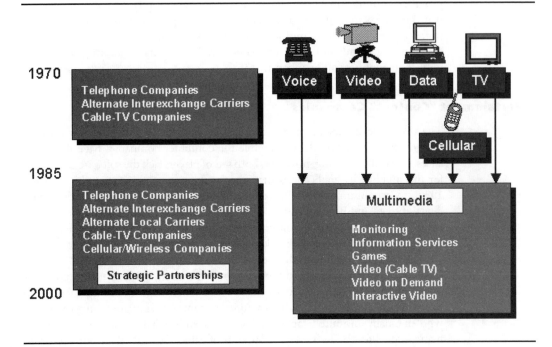

FIGURE 12.3   Merging of markets.

Cable TV Companies

Cable TV companies have been continuously upgrading their networks. To increase capacity, reliability, and picture quality, they have converted trunks from coaxial cable to optical fiber. Cable TV networks are now being transformed into multiservice broadband networks, using the headend (i.e., central receiver) as an authoring area for delivering integrated digital services by satellite or terrestrial means to the residence. The local loop portion of cable TV networks uses optical fiber platforms and incorporates network redundancy similar to that provided by telephone company SONET rings.

Computer Companies

For many computer companies multimedia represents an opportunity to reverse their declining revenues from mainframes and minicomputers and to finally enter the data communications business. Today, globally, there are hundreds of millions of personal computers and the numbers continue to grow because of unparalleled increases in low-cost processing power. (Worldwide sales of Intel X86 PCs are projected by International Data Corp. to exceed 60 million units in 1997. Companies are using their knowledge of PCs to become providers of multimedia servers and access appliances, relying upon their hardware and software expertise as well as their growing business in systems integration to provide custom multimedia-networked solutions. DEC, for one, is developing a broadband, interactive network linking manufacturing companies with their suppliers and subcontractors.

Entertainment "Content" Companies

The telephone companies and the cable TV companies own the networks for multimedia information delivery, while the computer companies have the appliances. Generally speaking, however, both sets of players lack the software or content for the multimedia applications that interest the general consumer. In the early 1990s, some consumer electronic companies did realize the benefits of delivering digital movies and began to purchase movie studios. Most recently, the cable TV companies that are interested in controlling the content as well as the delivery of movies-on-demand have also made overtures to entertainment companies.

Towards the middle of this decade, dozens of interactive TV alliances were (and continued to be) formed between entertainment, cable, phone, and computer companies with the focus on video-on-demand. Although there have been several market tests in which a digital system was simulated using operators who manually mounted tapes on VCRs, attention has turned toward automatic, digital test beds. Today, thousands of cable TV and telephone company customers are receiving trial services including movies-on-demand, interactive games, shopping, and distance learning.

Telecommunications Companies

The telephone company views multimedia as a group of services that will ride upon the existing public telephone network. Some of the envisioned features will allow users to

- Use a single device for access to all types of communications.
- Find the names of persons or businesses in an up-to-date worldwide directory and, with the touch of a single button, call or send E-mail, facsimiles, graphics, or video to them.
- Work at any office or home location equally effectively. Users will be able to send documents to other workers at anytime no matter where they are, then discuss and mark the shared work in real-time.
- Have transparent and easy access to thousands of information services.

For the near term, Bellcore has selected ISDN as the network access vehicle for their MediaCom service prototype (Figure 12.4). Their preferred user appli-

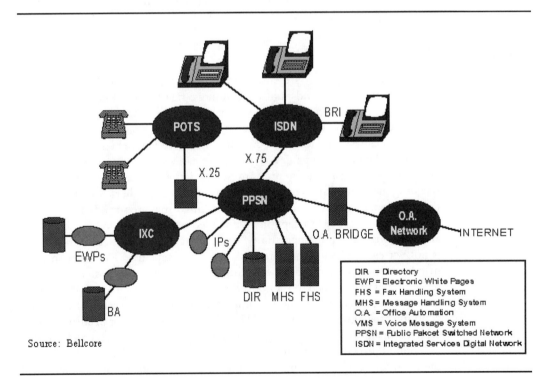

Source: Bellcore

FIGURE 12.4   Telephone company multimedia network.

TABLE 12.2   Video-on-Demand Network Requirement

Type	Rate (Mb/s)	Storage for two-hour movie in gigabytes	U.S.-wide storage for two-hour movie in terabytes
MPEG 1: VCR quality	1.5	1.4	1,400
MPEG 2: Broadcast quality	4–10	160	3,600–9,000
NTSC: Standard TV	177	160	160,000
HDTV	500	450	450,000

ance is the common personal computer, connecting to a packet data network. Attached to the packet network are remote servers that contain databases of electronic white pages, corporate directories, mail store-and-forward messages, fax handlers, and voice mail systems. The choice of ISDN, a relatively low-speed network, precludes many of the potentially lucrative video services such as video-on-demand that not only use high bandwidths but also require high-capacity servers (Table 12.2).

STRATEGIC ALIGNMENT

A wide variety of players are entering the multimedia field and filling various niches. Movie studios and publishing houses, traditional content developers, and software companies are joining with consumer electronics, telephone, cable TV, and computer companies to compete as multimedia providers. Many computer companies have formed education and entertainment divisions, while cable and telephone companies have forged alliances for control of the transport as well as the interactive content. At present, it is the network providers who are spending the most money on digital convergence. The others will invest when the broadband network transport and switching issues are sorted out.

The convergence of telephony with video and information services creates an enormous marketing opportunity for all of these companies. Nevertheless, it is not yet clear which companies will provide interactive TV, video-on-demand, video dialtone, and other information services, nor is it clear which technologies will be used to transport these services. Equipment company marketing decisions will become less important over time because, if the trend continues, telephone companies will continue to join together in alliances with cable companies. The networks they establish will provide access and impetus for content providers, software companies, the silicon suppliers who will build the chips for the TV set-top boxes, and the cable vendor companies who will build and sell the set-top boxes. Once the network infrastructure is firm, the content providers will have a reliable medium over which to send their products into millions of homes and businesses. The development of the next generation of communications will be an evolutionary process that will be driven by consumer needs and willingness to pay.

CONCLUSION

The telephone companies and cable television operators are competing for supremacy on the broadband digital highway local loop. Although their vision of interactive services is decades old, it was frustrated by the lack of bandwidth and processing power. Now, enabling multimedia technologies can rejuvenate aging public telephone networks and unidirectional cable TV networks, creating new and powerful complex interconnections that can carry interactive television, videoconferencing, and other innovative forms of audio and video bit traffic. Once they are generally available, these networks will motivate content and equipment providers to join. Until then, there are still some questions to be answered. Who will dominate the information highway? Will convergence occur naturally or via the takeovers of cable TV businesses by larger telecommunications companies? Not too surprisingly, the answers may well depend more on political and economic issues than on technical ones.

Broadband Network Standards

NORTH AMERICAN STANDARDS

ANSI Standards

ATM Standards

T1S1.5/92–001:	AAL SSCOP baseline document
T1.ATM–199X:	ATM layer functionality and specification
T1.AL4–199X:	AAL 3/4 common part T1.CBR-199X AAL for constant bit-rate services functionality and services
T1S1.5/92–005:	connectionless service layer functionality and services
T1S1.5/92–010:	AAL5 common part functionality and services
T1S1.5/92–111:	constant bit-rate AAL architecture references

SONET Standards

Phase I

T1.105:

- Byte-interleaved multiplexing format
- Line rates for STS—1, 3, 9, 12, 18, 24, 36, and 48
- Mappings for DS0, DS1, DS2, DS3
- Monitoring mechanisms for section, line, and path structures
- 192 kb/s and 576 kb/s DCC

T1.106: optical parameters for the long-reach single-mode fiber cable systems.

Phase II

T1.105R1:

- SONET format clarification and enhancements
- Timing and synchronization enhancements
- *Automatic protection switching* (APS)
- Seven-layer protocol stack for DCC and embedded operations channels
- Mapping of DS4 (139 Mb/s) signal into STS-3c

T1.117:   optical parameters for short-haul (less than two kilometers) multimode, fiber cable systems.

T1.102–199X: electrical specifications for STS-1 and STS-3 signals.

Phase III

- **T1.105.01–1994:** requirements for two-fiber and four-fiber bidirectional line-switched SONET rings.
- **T1.105.05–1994:** requirements for Tandem Connection Overhead layer for SONET.
- **T1.105.03–1994:** jitter requirements for SONET interfaces and at interfaces between SONET and asynchronous networks.
- **T1.119–1994:** management specification for SONET network elements using an OSI-compliant interface.

Bellcore—technical advisories and requirements for BOCs

ATM Standards

TA-NWT-001113:   asynchronous transfer mode (ATM) and ATM adaptation layer (AAL) protocols generic requirements

BISDN Standards

FA-NWT-001109:   broadband ISDN transport network elements framework generic criteria

FA-NWT-001110:   broadband ISDN switching system framework generic criteria

FA-NWT-001111:   broadband ISDN access signaling framework generic criteria for Class II equipment

TA-NWT-001112:   broadband-ISDN used to network interface and network node interface physical layer generic criteria

SR-NWT-001763:   preliminary report on broadband ISDN transfer protocols

SMDS Standards

TR-TSY-000772:	generic requirements in support of switched multimegabit data service
TR-TSY-000773:	local access switching system generic requirements in support of SMDS

SONET Standards

TR-TSY-000233:	wideband and broadband digital cross–connect systems generic requirements and objectives
TA-TSY-000253:	SONET transport systems: common generic criteria
TA-TSY-000303:	IDLC system generic requirements, objectives, and interface: feature set C—SONET interface
TR-TSY-000418:	generic reliability assurance requirements for fiber-optic transport systems
TA-TSY-000496:	SONET ADM generic requirements
TR-TSY-000496:	SONET add/drop multiplex equipment (SONET ADM) generic requirements and objectives
TR-TSY-000499:	transport systems generic requirements (TSGR): common requirements
TA-TSY-000755:	SONET fiber-optic transmission systems requirements and objectives
TA-TSY-000773:	local access system generic requirements, objectives, and interface in support of SMDS
TR-TSY-000782:	SONET digital switch trunk interface criteria
TA-TSY-000842:	generic requirements for SONET-compatible digital radio
TA-TSY-000917:	SONET regenerator generic requirements
TA-TSY-001042:	generic requirements for operations interfaces using OSI tools: SONET transport

ATM Forum

Q.2110/Q.2130:	ATM User-Network Interface (UNI) data-link protocol
Q.2931/Q2971:	ATM User-Network Interface (UNI) signaling protocol

IEEE—802.X series of LAN standards

IEEE 802.3:	Ethernet
IEEE 802.5:	token ring

IEEE 802.6:   distributed queue dual-bus subnetwork of a metropolitan area network

EUROPEAN STANDARDS

BISDN Standards

I.113:   vocabulary of terms for broadband aspects of ISDN

I.121:   broadband aspects of ISDN

I.150:   B-ISDN asynchronous transfer mode functional characteristics

I.211:   B-ISDN service aspects

I.311:   B-ISDN general network aspects

I.321:   B-ISDN protocol reference model and applications

I.327:   B-ISDN functional architecture

I.361:   B-ISDN ATM layer specification

I.362:   B-ISDN ATM adaptation layer (AAL) functional description

I.363:   B-ISDN ATM adaptation layer (AAL) specification

I.413:   B-ISDN user-network interface

I.432:   B-ISDN user-network interface—physical layer specification

I.610:   OAM principles of the B-ISDN access

OBTAINING STANDARDS

Standards documents in draft form may be obtained from the responsible committee. Published standards may be purchased by mail.

American Standards Institute (ANSI):

ANSI
Customer Service
11 W. 42nd Street
New York, NY 10036
Telephone: 212 642-4900
Facsimile: 212 302-1286

Bell Communications Research

Bellcore
331 Newman Springs Road
Post Office Box 7020
Red Bank, New Jersey 07701-7020
Telephone: 908 699-5800

International Electrotechnical Commission (IEC):

IEC Sales Dept.
Central Office of the IEC
3 rue de Varembe
PO Box 131
1211 Geneva 20
Switzerland
Telephone: 41 22 734 0150
Facsimile: 41 22 733 3843

International Telecommunication Union (ITU):
(Formerly CCITT)

ITU General Secretariat
Sales Section
Place des Nations
1211 Geneva 20
Switzerland
Telephone: 41 22 730 5111
Facsimile: 41 22 730 5194

Telecommunications Industry Association (TIA):

TIA
2500 Wilson Boulevard
Suite 300
Arlington, VA 22201
Telephone: 703 907-7700
Facsimile: 703 907-7727

2

Glossary

10BASE2 An 802.3 standard: 10 Mb/s transmission BASEband with 185 meters per thin (RG-58A/U) coaxial segment. Standard physical layer option for CSMA/CD.

10BASE5 An 802.3 standard: 10 Mb/s transmission BASEband with 500 meters per coaxial segment. Standard physical layer option for CSMA/CD.

10BASE-T An 802.3 standard: 10 Mb/s transmission BASEband over twisted-pair. Standard physical layer option for CSMA/CD.

AAL ATM Adaptation Layer: The ATM standards that specify the procedures to be followed to segment variable-length data packets into cells for transport through an ATM network and then to reassemble as they exit the network. The AAL is subdivided into the SAR and the convergence sublayer.

ANSI American National Standards Institute: United States' representative to the CCITT.

ATM Asynchronous transfer mode: A form of fast packet-switching technology, using an asynchronous time division multiplexing technique: The multiplexed information flow is organized into fixed blocks called *cells*. It is asynchronous in the sense that the recurrence of cells containing information from an individual user is not necessarily periodic.

Bandwidth A measure of information-carrying capacity.

BISDN Broadband ISDN. *See* ISDN

Bridge A device that connects two or more LAN networks and forwards frames between them. Bridges can usually be made to filter frames, for example, to forward certain traffic only.

Broadcast A packet delivery system that delivers a copy of a given packet to all hosts that are attached to it, is said to broadcast the packet.

CAD/CAM Computer-aided design/computer-aided manufacturing.

CCITT Comite Consulatif International Telegraphique et Telephonique, also known as the International Telegraph and Telephone Consultative Committee. Formerly a unit of the International Telecommunications Union (ITU) of the United Nations.

Cell A short, fixed-length packet used in the ATM high-speed packet-switching technique, consisting of 53 bytes, 5 of header and 48 of payload.

Cell Relay *See* ATM.

CLP Cell Loss Priority: A cell header field that is used to provide guidance to the network in the event of congestion.

Connectionless Communication without a path for end-to-end connection being established first. Sometimes called *datagram*. A LAN would be an example.

Connection-oriented Communication proceeds through three well-defined phases: connection establishment, data transfer, and connection release. Examples are X.25, Internet TCP, OSI TP4, and ordinary telephone calls.

CPE Customer premises equipment: Generic name for transmission devices that are located in, and owned by, the public service customers.

CRC Cyclic redundancy check: An error-detection scheme.

CSMA/CD Carrier sense multiple access with collision detection. A contended-access method in which stations listen before transmission, send a packet, and then free the line for other stations. Also the access method used in Ethernet and IEEE 802.3.

CSU Channel service unit: A digital DCE unit for DDS lines.

Datagram An abbreviated, connectionless, single-packet message from one station to another.

DCE Data communications equipment, or data circuit-terminating equipment. In common usage, synonymous with *modem*: The equipment that provides the functions required to establish, maintain, and terminate a connection, as well as the signal conversion required for communications between the DTE and the telephone line or data circuit.

DCS Digital cross-connect system. A programmable, electronic routing device deployed within a carrier's network to establish temporary end-to-end routes for high-capacity circuits.

DMA Direct memory access: A fast method of moving data between two processor subsystems without processor intervention.

DQDB Distributed queue dual bus: IEEE 802.6-defined cell-relay standard for SMDS.

DSU Digital service unit: DCE common equipment used to connect a customer's DTE to public network facilities or to CSU-equipped DS1 facilities.

DTE Data terminal equipment: The equipment serving the network DCE as the data source, the data sink, or both.

EDI Electronic data interchange: The computer-to-computer exchange of documents between different companies, using the telephone network.

EIA Electronics Industry Association: A standards group within ANSI for the electronics industry.

Ethernet An IEEE 802.3 LAN first developed by Xerox, then sponsored by DEC, Intel, and Xerox. An Ethernet LAN uses coaxial cables and CSMA/CD.

FDDI Fiber-distributed data interface: A high-speed networking standard. The underlying medium is fiber optics and the topology is a dual-attached, counter-rotating token ring.

Flow control Control of the rate at which hosts or gateways inject packets into a network or internet, usually to avoid congestion.

Frame Variable-length, addressed data unit identified by a label at layer 2 of the OSI reference model.

Frame relay A networking packet-delivery interface with an historical basis in X.25.

Gateway A special-purpose, dedicated computer that attaches to two or more networks with different communications protocols and routes packets from one to the other.

GFC Generic flow control: A cell header field used for multiplexing for access to an ATM network.

Groom and Fill Lower rate signals are taken from incoming signals and added to outgoing higher-rate signals.

HEC Header error control: A cell header CRC field that can be used to correct single-bit errors in the header and to detect multiple-bit errors.

Header The bits within a cell allocated for functions required to transfer the cell payload within the network.

IEEE The Institute of Electrical and Electronic Engineers is an engineering association active in defining LAN standards.

Internet A collection of packet-switching networks interconnected by gateways with protocols that allow them to function logically as a single, large, virtual network. When written as **Internet**, the word refers specifically to the Defense Advanced Projects Research Agency (DARPA) Internet and the TCP/IP protocols.

IP Internet Protocol. *See* TCP/IP

ISDN Integrated digital services network: ISDN combines voice and digital network services in a single medium, making it possible to offer customers digital data services as well as voice connections through a single wire. The standards that define ISDN are specified by ITU.

ISO International Standards Organization: An international body that drafts, discusses, proposes, and specifies standards for network protocols. ISO is best known for its seven-layer reference model that describes the conceptual organization of protocols.

Isochronous An isochronous service delivers bandwidth at regular, predictable intervals.

ITU International Telecommunications Union: The United Nations standards group responsible for telecommunications.

Jitter Uncertainty in signal location due to mechanical or electrical changes.

LAN Local area network: A data communications network confined to a limited geographic area (up to six miles, or about 10 kilometers) with moderate to high data rates (100 Kb/s to 100 Mb/s). The area served may consist of a single building, a cluster of buildings, or a campus-type arrangement.

LATA Local access and transport area. The geographic area over which a local telephone company can offer service without having to access another carrier.

Latency The amount of time elapsed between the transmission and reception of a signal.

MAC Media access control.

MAN Metropolitan area network.

MAU Medium access unit: A transceiver for 802.3 10BASE5 and 10BASE2.

Message A complete transmission: Used as a synonym for a *packet* or *frame* of information, but often made up of several packets.

MIB Management information base: A collection of objects that can be accessed via a network management protocol.

Multicast A technique that allows copies of a single packet or cell to be passed to a selected subset of all possible destinations.

NISDN Narrowband or national ISDN.

NIC Network interface controller: Circuitry, usually a PC expansion card, that connects a workstation to a network.

Object Reusable software modules that include both data such as customer profile information and procedures—for example, how that information should be used.

OSI Open systems interconnection: Refers to a seven-layer hierarchical reference structure developed by the ISO for defining, specifying, and relating communications protocols. In the OSI model, suites of communications protocols are arranged in layers.

Packet An information block identified by a label at layer 3 of the OSI reference model. It is the unit of data sent across a packet-switching network. *See* frame and *PDU*

PC Personal computer: For mainframe people this is the devil's tool, but for the masses it is the affordable computer of choice. The PC is highly integrated consisting of a microprocessor, RAM memory, and hard disk.

PCM Pulse-code modulation: A modulation technique used to convert analog voice signals into digital form. Used for voice multiplexing on T1 circuits.

PDU Protocol data unit: OSI terminology for *packet*. A PDU is a data object exchanged by protocol machines (entities) within a given layer. *See* frame and packet

POTS Plain old telephone service: Transported via PSTN, POTS offers traditional two-way voice conversation.

Protocol A formal description of message formats and the rules two or more machines must follow to exchange those messages.

PSTN Public-switched telephone network.

PT Payload type: A cell header field indicating the type of information in the cell payload.

PTT Post, Telephone, and Telegraph Authority: The government agency that functions as the communications common carrier and administrator in many areas of the world.

PVC Permanent virtual circuit: In a packet-switched network, a fixed virtual circuit between two users; no call setup or clearing procedures are necessary. Contrast with *SVC*.

Repeater A hardware device that propagates electrical signals from one cable to another without making routing decisions or providing packet filtering. In OSI terminology, a repeater is a *physical layer intermediate system. See* bridge, gateway, and router

Router A system responsible for making decisions about which of several paths network traffic will follow. To do this, it uses a routing protocol to gain information about the network and algorithms to choose the best route based on predetermined criteria. In OSI terminology, a router is a network layer intermediate system. *See* bridge, gateway, and repeater.

SAR Segmentation and reassembly. One of two sublayers of the AAL with the functions of—on the transmitting side—the segmentation of higher-layer PDUs into a suitable size for the information field of the ATM cell and—on the receiving side—the reassembly of the particular information fields into higher-layer PDUs.

SDH Synchronous digital hierarchy: Europe's version of SONET.

Server A variety of computer devices ranging from fast personal computers with large amounts of memory to data switches.

SMDS Switched multimegabit data service: A public packet-switching service proposed by Bellcore and offered by the telephone companies.

SNMP Simple network management protocol: The network management protocol of choice for TCP/IP-based internets.

SONET Synchronous optical NETwork: An advanced, fiber-based public network defined by a large family of related technical standards. *See* SDH

STM Synchronous transfer mode: A transfer mode that offers, periodically to each connection, a fixed-length word. Contrast with *ATM*.

SVC Switched virtual circuit: In a packet-switched network, a temporary virtual circuit between two users. Contrast with *PVC*.

TCP/IP Transport control protocol/Internet protocol. TCP is the protocol that provides reliable, end-to-end stream transport. IP is the universal protocol of the Internet that defines the unit of transfer to be the IP datagram and provides the universal addressing scheme for hosts and gateways.

TDM Time division multiplexing: A technique used to multiplex multiple signals onto a single hardware transmission channel by allowing each signal to use the channel for a short time before going on to the next.

Transceiver A radio frequency device that serves as both a transmitter and receiver.

UTP Unshielded twisted-pair: One or more pairs of twisted insulated copper within a single plastic sheath cable.

VCI Virtual channel identifier: A routing field in the header of a cell. Used to identify the virtual connection to which the cell belongs.

VPI Virtual path identifier: A routing field in the header of a cell.

WAN Wide-area network.

Workstation A computer that is usually more powerful than a PC, but less than a minicomputer that fits on a desk.

X.25 X.25 is a CCITT standard for packet-switching that was approved in 1976. Its significance was that it standardized the structure of the packets and codified the functionality that could be embedded into them to facilitate networking.

Acronyms

AAL ATM adaptation layer

ADM add/drop multiplexer

ADPCM adaptive PCM

AIN advanced intelligent network

ALC analog loop carrier

AMPS advanced mobile phone system

ANI automatic number identification

ANSI American National Standards Institute

APS automatic protection switching

ARP address resolution protocol

ATM asynchronous transfer mode

AUI attached unit interface

BRI basic rate interface

CAD computer-aided design

CCITT Consultative Committee on International Telephony and Telegraphy, now ITU

CDDI copper distributed data interface

CDMA code division multiple access

CDPD Cellular Packet Radio Data

CEC Commission of European Communities

CIR committed information rate

CIX Commercial Internet Exchange Association

CLNP connectionless network protocol

CLP Cell loss priority

CMIP common management information protocol

CNRI Corporation for National Research Initiatives

CO central office

Codecs CODer/DEcoders

CPE customer premises equipment

CS convergence sublayer

CSMA/CA carrier sense, multiple access with collision avoidance

CSMA/CD carrier sense multiple access with collision detection

CSU/DSU channel service unit/data service unit

DARPA Defense Advanced Research Projects Agency

DASs dual-access stations

DCC data communications channel

DCS digital cross-connect system

DDS digital dataphone service

DE discard eligibility bit

DLCI data-link connection identifier

DoD Department of Defense

DPA demand priority access

DQDB distributed queue dual bus

DSX digital signal cross-connect frame

E-mail electronic mail

ECSA Exchange Carriers Standards Association

EDI electronic data interchange

EMI electromechanical interference

EOC embedded operations channel

ETSI European Technical Standards Institute

FDDI fiber distributed data interface

FDMA frequency division multiple access

FECN/BECN forward and backward explicit congestion notification

FITL fiber in the loop

FM frequency modulation

FOTs fiber-optic terminals

FRAD frame relay access device

FSK frequency shift-keying

Gb/s gigabits per second

GFC generic flow control

GGP gateway/gateway protocol

GUI graphical user interfaces

HEC Header error control

HIPPI high-performance parallel interface

HLDC high-level data-link control

HLI high-speed LAN interconnection service

HRC hybrid ring control

HSSI high-speed serial channel interface

I/O input/output

IDLC integrated digital loop carrier

IEEE Institute of Electrical and Electronic Engineers

IETF Internet Engineering Task Force

IS information system

ISDN integrated service digital network

ISO International Standards Organization

IT Information technology

ITU International Telecommunications Union

IVR interactive voice response

IXCs interexchange carriers

LANs local area networks

LATA local access and transport area

LEC local exchange carrier

LLC logical link control

LTE line-terminating equipment

MAC media access control

MANs metropolitan area networks

MAU media access unit

MDF main distribution frame

MIB management information base

MII media independent interface

MIPS millions of instructions per second

MIT Massachusetts Institute of Technology

MOU minutes of use

MPEG motion pictures expert group

MVC multicast virtual circuit

NLM NetWare loadable module

NEXT near-end crosstalk

NIC network interface card

NISDN narrowband integrated services digital network

NNI network-to-network interface

NRZ Non-return to zero

NSF National Science Foundation

OAM&P operations, administration, maintenance, and provisioning

OC optical carrier

ONU optical network unit

OSPF open shortest path first

OSI open systems interconnect

OSS operations support system

PBX private branch exchange

PCM pulse code modulation

PCS personal communication service

PDA personal digital assistant

PDU protocol data unit

PMD physical medium-dependent

POH path overhead

POP point-of-presence

POTS plain old telephone service

PPS path protection switching

PRC primary reference clocks

PRI primary rate interface

PSTN public-switched telephone network

PT payload type

PTE path terminating equipment

PTM packet transfer mode

PTT private telephone and telegraph company

PUC Public Utilities Commission

PVC permanent virtual circuit

RAM random access memory

RBOC Regional Bell Operating Company

RDT remote digital terminal

RFI radio frequency interference

RFT remote fiber terminal

RIP routing information protocol

RISC reduced instruction set computing

SAIC Science Applications International Corp.

SAR segmentation and reassembly

SAS single access station

SDH synchronous digital hierarchy

SDLC synchronous data-link control

SIP SMDS interface protocol

SMDS switched multimegabit data service

SMT station management transcending

SMTP simple mail transfer protocol

SNA system network architecture

SNI subscriber network interface

SNMP simple network management protocol

SONET in North America, SDH in Europe synchronous optical network

SPE synchronous payload envelope

ST1 stratum 1

STE section terminating equipment

STM synchronous transport mode

STP standard twisted-pair

STS synchronous transport signal

SVC switched virtual circuit

TCNS Thomas-Conrad Network System

TCP/IP transmission control protocol/internet protocol

TDM time division multiplexer

TDMA time division multiple access

Telco Telephone company

TLS transparent LAN service

TM terminal multiplexer

TOH transport overhead

UDLC universal digital loop carriers

UMMC University of Maryland School of Medicine

UNI user-to-network interface

UTP unshielded twisted-pair

VCI virtual channel identifier

VPI virtual path identifier

VT virtual tributary

WAN wide-area network

About the Author

Robert P. Davidson, Ph.D. is president of Reference Point, a provider of computer-based training (CBT) and seminars in communications. He is the author of six books on data communications, cable TV, and telephony, as well as numerous multimedia CBT. He has over 20 years of experience in telecommunications engineering, marketing, and sales with AT&T, Bell Telephone Laboratories, and General DataComm. He is a Senior Member of the IEEE and has been named to the IEEE Computer Society List of Leaders. He is also a past member of the editorial board of *Spectrum* magazine and of the International Test Conference. Robert can be reached via E-mail at RPDINC@aol.com.

Index